TAXONOMY OF FINFISH

Theory Manual

THE AUTHORS

N. Karthik is an Assistant Professor in the Department of Fisheries Resource Management at the College of Fisheries Science, St. Devasahayam Institute of Fisheries Science and Technology, Midalam, Kanniyakumari dist., Tamil Nadu. He graduated with a Bachelor's degree in Fisheries Science from Dr MGR. Fisheries College and Research Institute, Ponneri, Tamil Nadu. He had his Master's degree in Fisheries Science from the Faculty of Fisheries, Sher-E-Kashmir University of Agricultural Sciences and Technology. He has qualified for ASRB – NET (National Teaching Eligibility Test) in Fisheries Resource Management (2023) and cracked the ICAR-JRF examination (2021). During his professional tenure, he diligently collected over 100 different species of finfish from the entire Kanyakumari coast, preserving them at the "Mukkadal (Tri-seas) Marine Museum" in the College of Fisheries Science, DIFST. Additionally, he has successfully conducted a capacity-building programme with funding obtained from NFDB, Hyderabad. Currently, he is leading an institutional research project.

S. Shamini is an Assistant Professor in the Department of Fisheries Resource Management at the College of Fisheries Science, St. Devasahayam Institute of Fisheries Science and Technology, Midalam, Kanniyakumari dist, Tamil Nadu. She graduated with a Bachelor's degree in Fisheries Science from Dr MGR. Fisheries College and Research Institute, Ponneri, Tamilnadu and has a Master's degree in Department of Fisheries Biology and Resource Management at Fisheries College and Research Institute in Thoothukudi, Tamil Nadu. During her professional tenure, she successfully conducted a capacity-building programme with funding obtained from NFDB, Hyderabad.

Prof. S. Felix is a distinguished figure in the field of fisheries education and technology consulting, with a career spanning over four decades. Currently serving as the Dean of the College of Fisheries Science at DIFST (Devasahayam Institute of Fisheries Science and Technology) in Midalam, Kanniyakumari district, Tamil Nadu and also President of AQUAFIN (Aquaculture Technology & Research Foundation India), he has held several prestigious positions, including Vice Chancellor of Tamil Nadu Fisheries University (TNJFU) and President of the World Aquaculture Society (WAS)-APC. Prof. Felix is recognized for his exemplary leadership and contributions to the fisheries and aquaculture sector, demonstrated through his involvement in various committees and organizations. He is a prolific author, having published numerous research papers, books, and manuals, and has played a pivotal role in introducing new fisheries science courses and establishing fisheries colleges and research centres.

TAXONOMY OF FINFISH
Theory Manual

— *Authors* —

N. Karthik

S. Shamini

S. Felix

College of Fisheries Science
St. Devasahayam Institute of
Fisheries Science & Technology
Midalam, Kanniyakumari Dist.629193 (Tamil Nadu)

2025

Daya Publishing House®
A Division of
Astral International Pvt. Ltd.
New Delhi – 110 002

ISBN: 978-93-5919-570-4

Published by : **Daya Publishing House®**
A Division of
Astral International Pvt. Ltd.
– ISO 9001:2015 Certified Company –
4736/23, Ansari Road, Darya Ganj
New Delhi-110 002
Ph. 011-43549197, 23278134
E-mail: info@astralint.com
Website: www.astralint.com

Foreword

The taxonomy of finfish offers a comprehensive exploration of the diverse array of aquatic finfish species inhabiting Indian waters. Beginning with an introduction to the principles of taxonomy and nomenclature, the manual progresses to dissect the morphological, morphometric and meristic characteristics that form the basis of finfish taxonomy. From the ancient Ostracoderms to the modern Teleostomes, each lecture of this manual delves into the unique traits and evolutionary adaptations defining various taxa of the classes Chondrichthyes and Actinopterygii. Commercially important marine and freshwater fish orders such as Clupeiformes, Perciformes and Cypriniformes are meticulously examined, shedding light on their economic significance and ecological roles within Indian ecosystems. Further, the taxonomy extends beyond finfish to encompass marine mammals and reptiles, providing a holistic understanding of the interconnected food web in Indian waters. This comprehensive manual serves as an invaluable resource for scholars, researchers, and conservationists alike, fostering a deeper appreciation for the rich biodiversity of Indian finfish.

It is with great pleasure that I recommend this well-planned manual on Finfish Taxonomy and Fish Identification to all the beneficiaries viz. students, staff, researchers, and conservationists as it is tailored to suit the needs of learners in our diverse Indian context. I also sincerely believe that this manual would greatly help all the beneficiaries in acquiring adequate knowledge on Taxonomy of Finfish.

I extend my best wishes to all those who have contributed in the creation of this useful publication.

Prof. Dr. Ramasamy Santhanam
Ex-Dean, Fisheries College & Research Institute
Tamil Nadu Veterinary and Animal Sciences University
Thoothukudi, TN

Preface

Welcome to the comprehensive manual for 'Taxonomy of Finfish'. This manual serves as a foundational resource for students, researchers, and enthusiasts delving into the fascinating world of fish taxonomy. Structured into five units, each lecture is meticulously crafted to provide an in-depth exploration of key topics essential for understanding the taxonomy and identification of finfish species.

In Unit I, we begin with an exploration of the fundamental principles of taxonomy, followed by an examination of nomenclature and classification criteria. Unit II delves into the morphological, morphometric, and meristic characteristics that underpin taxonomic classification. Unit III focuses on the intricate characters and classifications of finfish, including the evolutionary history of ancient groups such as Ostracoderms and Placoderms, and the diverse morphology of Chondrichthyes, Sharks, and Batoid Fishes. Additionally, lectures on Teleostomes and Actinopterygii provide insights into the higher bony fishes. The subsequent units delve into the taxonomy of commercially important marine and freshwater fishes, including Clupeiformes, Perciformes, Pleuronectiforms, Siluriforms, and others. Furthermore, our exploration extends to marine mammals and reptiles, shedding light on their significance in marine ecosystems. Unit V concludes with an overview of modern taxonomical tools used in fish identification, equipping readers with practical skills for species recognition and classification.

We extend our sincere gratitude to all the stakeholders who have made this manual possible, and we hope it serves as a valuable companion in your journey through the captivating realm of Fish Taxonomy.

N. Karthik

S. Shamini

S. Felix

Contents

UNIT – I

Chapter-1

Principles of Taxonomy

Introduction

- **Taxonomy** is the science of systematically classifying and naming organisms based on their shared characteristics and evolutionary relationships, creating a structured framework for understanding biodiversity.
- It includes **classifying and arranging every known species of creature**, including all known animal, plant, fungal, and microbial species, into a hierarchical structure that represents their evolutionary links and common traits.
- **Carolus Linnaeus** is known as the **father of taxonomy**. He categorized the organisms based on their differences and similarities.
- The term **'Taxonomy'** is derived from the **Greek words'Taxis'** meaning **arrangement**, and '**nomos**,' meaning **law**. It was **first proposed by Candolle in 1813**.
- The **primary aim of taxonomy** is to **provide an organised and systematic framework for understanding and expressing the diversity of life on Earth**. In taxonomy, there are various levels of hierarchy, and the Linnaean method of classification is widely used.

Fish have been on Earth	About 450 million years.
Total number of recognized vertebrate species	65,000 to 70,000 (24,618 valid fish species).
Currently Living Species	Approximately28,500.
Total Recognized fish families	482 (8 largest families, each boasting over 400 species, collectively account for approximately 33% of all species). These families, listed in descending order of species count, are: Cyprinidae Gobiidae Cichlidae Characidae Loricariidae Labridae Belitoridae Serranidae

Classification: Organizing organisms into groups or categories based on shared characteristics and evolutionary relationships.

Identification: The process of placing an individual organism into a specific taxonomic category using a deductive procedure.

Taxa (Taxon): Groups of organisms recognized in a classification and given biological names, such as orders (*e.g.,* Clupeiformes), families (*e.g.,* Clupeidae), and genera (*e.g.,* Sardinella).

Category: The level or rank at which a taxon is placed, such as orders, families, and genera.

Systematics: It is a biological science that involves the discovery of names, determination of relationships, classification, and the study of the evolution of living organisms.

Fish Taxonomy: It is a field of study that involves the classification of fish based on their similarities and differences, following internationally accepted principles, rules, and regulations.

Three Types (Stages) of Taxonomy

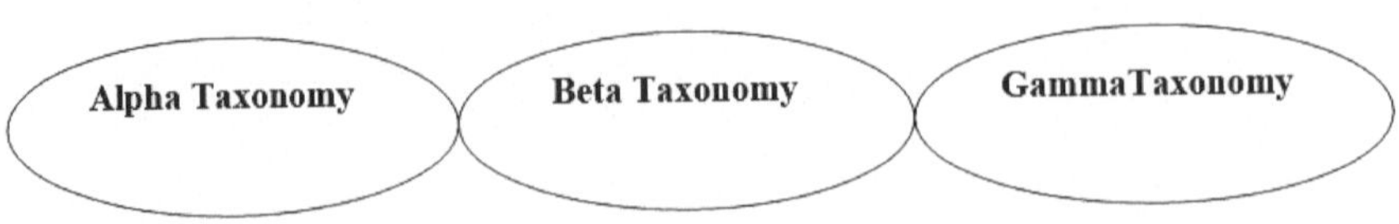

- **Alpha Taxonomy**: This stage involves describing new species and placing them within appropriate genera.
- **Beta Taxonomy**: At this level, relationships among species and higher categories (like genera, families, *etc.*) are determined.
- **Gamma Taxonomy**: Gamma taxonomy focuses on studying variations within species (intraspecific variations) and understanding their evolutionary relationships.

Importance of Taxonomy

- Taxonomy organizes and names the vast diversity of life.
- It aids species identification and conservation.
- Taxonomy informs evolutionary and ecological studies.
- It plays a role in agriculture, medicine, and environmental monitoring.
- Taxonomy is foundational to understanding life's evolutionary history.

Conclusion

A key component of biology is taxonomy, which provides a methodical framework for categorizing biological organisms. Taxonomy enhances our understanding of biodiversity by identifying common characteristics and evolutionary linkages. It also serves as a basis for other scientific disciplines, from species identification and conservation to evolutionary study and environmental monitoring. It supports our understanding of the natural world and its numerous interconnections, and it is an essential tool for both research and practical applications.

Chapter-2

Nomenclature and Types

Introduction

The field of biology encompasses a massive and intricate system of organisms, each having a unique combination of characteristics and features. Scientists have established a systematic method for naming, classifying, and recognizing these creatures in order to make sense of their complexity. This method, which has its origins in taxonomy and systematics, is very important in assisting us in understanding the broad spectrum of life on Earth.

ICZN (International Commission on Zoological Nomenclature)

- The International Code of Zoological Nomenclature (ICZN) is a set of rules governing the naming and classification of animals.
- Its history began in the 19th century when there was confusion in zoological literature due to multiple naming systems and the growth of species knowledge. The need for uniformity in naming and stability led to early efforts like the British Association Code (Stricklandian Code).
- In 1889, the 1st International Congress of Zoology started the move towards an international code.
- The commission was set up in 1895. It consists of 25 members from 20 countries. It operates in two main ways.
- First, it publishes the International Code of Zoological Nomenclature containing the rule universally accepted as governing the application of scientific names to all organisms which are treated as animals.

- Secondly, it gives rulings on individual nomenclatural problems brought to its attention, so as to achieve internationally acceptable solutions.
- Several million species of animals are recognized and more than 2000 new generic names and 15000 new specific names are added to the zoological literature every year. With such a multiplicity of names, problems are bound to occur.
- The commission operates through its quarterly journal, the Bulletin of Zoological Nomenclature, in which problems needing a formal decision by the commission are published for discussion by the zoological community., which later became the International Commission on Zoological Nomenclature.
- The commission is under the auspices of the International Union of Biological Sciences (IUBS). The American Association for Zoological Nomenclature and The European Association for Zoological Nomenclature facilitate liaison between zoologists and the ICZN and provide financial support.
- The first formal Code was published in 1905. Several editions and revisions followed.
- In 1972, the responsibility for the Code was transferred to the International Union of Biological Sciences (IUBS). The third edition was published in 1985, considering advancements in taxonomy and technology.
- An Editorial Committee was formed in 1988, leading to the development of a fourth edition of the Code. It was based on extensive discussions and public input, with the final draft being accepted in 1997.
- Throughout its history, the ICZN has aimed to provide stability and consistency in the naming and classification of animals, adapting to the evolving needs of the scientific community.

Nomenclature

- Nomenclature, in biological classification, system of naming organisms. The species to which the organism belongs is indicated by two words, the genus and species names, which are Latinized words derived from various sources.
- This system, which is called the Linnaean system of binomial nomenclature, was established in the 1750s by Carolus Linnaeus.
- Subsequent to the work of Linnaeus, a proliferation of binomial names took place as new species were established and higher taxonomic categories were formed, with the result that by the late 19th century there was much confusion in the nomenclature of many groups of organisms.
- In the 20th century, the establishment of rules by international committees in the fields of zoology, botany, bacteriology, and virology has done much to clarify the situation.

- Contrary to the widely held view that scientific names, once assigned, are fixed and universal in their use, continuing research on the relationships of organisms and probing into the history of names, coupled with disagreements among scientists on the validity of certain names, results in multiple names being applied to some well-known species. The international rules, however, are gradually bringing stability to the taxonomy of many groups through the minimizing of name changes, the use of standard methods of establishing new names, and the functioning of respected committees to arbitrate controversies.

Zoological Nomenclature and Finfish Taxonomy

Zoological nomenclature is a system used for assigning scientific names to different taxonomic units of animals, encompassing both currently existing and extinct species. In the realm of finfish taxonomy, specific rules must be rigorously adhered to in order to establish the validity of a taxonomic unit.

The Linnaean Hierarchy

In the classification of finfishes, taxonomists utilize the Linnaean hierarchy, a structured system of categorical ranks that applies to various taxonomic groups. Within this hierarchical structure, each rank, except for the lowest one, includes one or more subordinate categories. The widely accepted categories within this system are as follows:

1.	Kingdom
2.	Phylum
3.	Sub phylum
4.	Class
5.	Sub class
6.	Super order
7.	Order
8.	Sub order
9.	Super family
10.	Family
11.	Subfamily
12.	Genus
13.	Sub genus
14.	Species
15.	Subspecies

Classification of Finfishes

In the classification of finfishes, a structured hierarchical system is employed. Typically, six categories are utilized: species, genus, family, order, class, and phylum. Within this hierarchy, the higher taxonomic categories encompass kingdom to genus levels, while species and subspecies are considered lower taxonomic units.

Standardized Naming Conventions

In the classification of finfish, specific suffixes are consistently applied to different taxonomic ranks. For instance, finfish orders typically conclude with the suffix 'formes,' such as Cypriniformes and Clupeiformes. Superfamilies adopt the standardized ending 'oidea,' as seen in Clupeoidea and Sepioidea. Families of finfish are identified by the ending 'idea,' exemplified by Scombridae and Sciaenidae, while tribes conclude with 'ini,' like Carangini.

Latin and Greek Influence

Binomial nomenclature, used for both animals and plants, draws extensively from Latin, especially in the naming of higher taxa like families and orders. These systematic names incorporate various adjectives and modifiers, with many originating from Latin, Greek, or local languages. While not all components of scientific names are Latin, they are treated grammatically as if they were. This approach involves adjusting word endings, such as -a becoming -ae and -us changing to -i to indicate possession. Furthermore, names derived from people's names add -i for males and -ii for females. This compilation of Latin and Greek terms frequently employed in systematic names aims to aid those unfamiliar with classical languages in comprehending and recalling scientific names of organisms.

Binominal Nomenclature

Binomial Nomenclature - Simplifying Species Naming

- ❑ Explanation of binomial nomenclature as a system for naming organisms with two terms: genus and species.
- ❑ Discussion of the two international codes: ICBN for plants and ICZN for animals. Ensuring a unique, globally recognized name for each organism.
- ❑ Components of Scientific Names - Explanation of the structure of scientific names, consisting of a generic name and a specific epithet. Overview of the conventions in writing these names.

Rules for Writing Scientific Names

- ❑ Details about the use of Latin in scientific names and the associated italics.
- ❑ Capitalization for the genus name and lowercase for the species name.
- ❑ How handwritten and typed names are represented to emphasize Latin origin.

Importance of Binomial Nomenclature

- ❑ The significance of binomial nomenclature in addressing the vast diversity of organisms globally.
- ❑ Its role in preventing confusion in identification and classification.

Drawbacks of Binomial Nomenclature

- Discussion of challenges, such as determining the correct name when multiple names are in use and the application of the law of priority.
- Recognition issues for names predating those in Linnaeus's "Systema Naturae."

Authorship in Scientific Naming

- The author or authors of a scientific name are the individuals who first formally establish the species by designating a type specimen.
- Author names are cited following the scientific names of the species they described (e.g., *Scomberoides commersonnianus* Lacepede, 1802, *Sardinella longiceps* Valenciennes).
- Citing the original author is not only a matter of giving credit but also serves to assign responsibility for the name and aids in locating the original description.
- In cases where a species group taxon is initially described in one genus and later transferred to another genus, the author(s) of the species group name should be enclosed in parentheses (e.g., *Arius sona* (Hamilton, 1822) = *Tachysurussona* (Hamilton, 1822); *Amblygastersirm* (Walbaum, 1792) = *Sardinella sirm* (Walbaum, 1792).
- If one desires to cite both the original author of the species group name and the reviser who transferred it to another genus, the reviser's name should follow in parentheses, enclosing the original author (e.g., *Scomberoides tala* (Cuvier) Smith Vaniz (1973).

Validity in Nomenclature

- Validity pertains to the rights of names concerning homonyms and synonyms.
- Synonyms are distinct names used for the same species, while homonyms are names with identical spellings that denote either different species group taxa within the same genus or objectively different taxa within the genus or family group.
- The earliest published synonym is termed the senior synonym, while later synonyms are junior synonyms.

Types of Synonyms

- Objective synonyms encompass names that objectively refer to the same entity, like a new name replacing a supposedly preoccupied one or names based on the same specimen or illustration.
- Subjective synonyms involve names based on different materials.

Emendations in Nomenclature

- Intentional Changes in Spelling- Emendations refer to intentional changes in the original spelling of a name. Two types of emendations exist:
- Justified emendation involves correcting an incorrect original spelling, and the emended name retains the date and authorship of the original spelling.
- Unjustified emendation occurs when the emended name is recognized in nomenclature with its own date and author.

Conclusion

Nomenclature is a foundational practice in science that provides standardized naming and classification, enabling clear communication and facilitating scientific research and education.

Chapter-3

Classification, Interrelationships and Criteria for Generic and Specific Identification

The main categories of Type Specimens, which are essential in the field of taxonomy for the proper identification and classification of organisms, include:

Holo type

The holotype is the **single specimen that the original author designates** as 'the type' at the time of publishing the initial description. It serves as the primary reference specimen for the species.

Syn type

In cases where there is no holotype, all specimens within the type series are referred to as syntypes. Syntypes may encompass specimens not personally examined by the author but were **based on previously published descriptions** or figures that contributed to the taxonomic definition.

Para type

Once the **holotype has been designated, the remaining specimens within the type series are labeled as 'paratypes.'** This differentiation helps in identifying the various components of the original type series.

Lecto type

The lectotype is chosen from a series of syntypes and is typically done by a **specialist during revision work**. It is not undertaken solely to add a type specimen to the collection. If a species description is clearly based on a specific specimen, that specimen is designated as the lectotype.

Neo type

When the original type specimens are known to be lost or were officially suppressed by a governing body, a neotype is selected from **subsequent specimens to serve as the new primary reference.**

Allo type

An allotype is a paratype of the **opposite sex to the holotype.** This distinction is particularly useful in species where sexual dimorphism plays a significant role in identification.

Topo type

A topotype is a **specimen collected from the same geographic location (type locality) as the original description**. It helps verify the consistency of the species within its known habitat.

Paratopo type or Iso type

An isotype is a specimen, **other than the holotype, collected from the same location as the holotype** and included in the initial species description. This further supports the species identification.

Nomenclature Changes in Taxonomy Involve Specific Terms

Nomen Nudum

This term is used to describe a species name that is published **without satisfying** the conditions for availability, rendering it unofficial and unaccepted in taxonomic records.

Nomen Dubium

A nomen dubium is a name given to a nominal species when the available **evidence is insufficient** to clearly identify the zoological species it was intended for. It indicates uncertainty in species classification.

Nomen Oblitum

A nomenoblitum refers to a name that has remained **unused as a senior synonym** in zoological literature for more than fifty years. Such names are considered forgotten and no longer in active use.

Nomen Conservandum

This term signifies a name that has been **preserved through the action of a commission** and officially placed on the appropriate official list. It is intended to ensure the continued use of certain taxonomic names for stability and clarity in scientific nomenclature.

Periods of Classification

1 st Period - The Ancient Odyssey (Early Naturalists)

- Think back to ancient Greece with the likes of Hippocrates and Aristotle. They were the early explorers, sketching out the first maps of animal classification based on observations of characteristics and habits.
- **John Ray**, a pioneer in his own right, ventured further into the **uncharted territories of classification** during his time. However, his system was like a sketch, not yet the masterpiece we know today.

2 nd Period - The Linnaean Revolution (Linnaean Era)

- Fast forward to the **18th century** and meet **Carl Linnaeus, the "Father of Taxonomy."** He was the **Copernicus of the biological world, introducing the binomial nomenclature and orbiting around Aristotle's principles**.
- Linnaeus provided the first well-structured taxonomy, offering a roadmap to the kingdom of life.

3rd Period - The Age of Exploration (Empirical Approach)

- Now, set sail through the 18th and 19th centuries. The world became smaller as explorers like **Lamarck and Cuvier** ventured into the unknown. This period was filled with subtle but significant developments.
- Zoologists crisscrossed the globe, **unveiling new creatures in the mysterious lands of Africa, Australia, and America**.

4th Period - The Darwinian Revolution (Darwin and Phylogeny)

- Arrive in the **19th century**, where **Charles Darwin** embarked on the voyage of a lifetime. He uncovered the secrets of evolution, forever changing the course of taxonomy.
- **Ernst Haeckel** added colorful branches to the tree of life with his concept of **"protista" and complex phylogenetic trees.**

5 th Period - The Varietal Symphony (Population Systematics)

- Leap into the **20th century** and witness the emergence of **population systematics**. This period was all about fine-tuning the notes within species.
- It **harmonized with classical taxonomy**, exploring the rich variations hidden within a single species.

6th Period - The Genomic Renaissance (Modern Taxonomy)

- Step into the present era of taxonomy. It's a time of reevaluation, where the entire classification theory is under the microscope.
- Here, we wield the tools of biochemistry and molecular markers, deciphering the genetic code to paint a more precise and colorful picture of the living world.

Criteria for Generic and Specific Identification

- Generic level fish species identification involves the classification and recognition of fish species at the genus or generic level. This means identifying and categorizing fish species into groups that share common characteristics and evolutionary relationships.
- Generic level identification keys are essential tools in taxonomy for assisting in the identification of organisms. They are designed to present diagnostic characteristics in a structured way, offering a series of choices that lead to the accurate determination of the taxon in question. Keys not only aid in identification but also play a crucial role in taxonomic analysis.

Purpose of Keys

- The primary purpose of identification keys is to facilitate the identification of a taxon. They serve as systematic guides to help users determine the identity of a particular genus or group of organisms.

Diagnostic Characters

- Taxonomists preparing keys must carefully select and evaluate the diagnostic taxonomic characters. These are the characteristics that most effectively distinguish one taxon from another.

Integral to Taxonomic Procedure

- Keys are an integral part of the taxonomic procedure and are used as a means of presenting taxonomic findings. They provide a structured pathway for users to navigate through the identification process.

Dichotomous Nature

- An effective key should be dichotomous, meaning it provides a sequence of paired choices, typically in a "yes or no" format. At each decision point, the key offers no more than two alternatives.

Clarity and Precision

- Each alternative presented in the key should be clear-cut and precise. Users should easily understand and select the appropriate option at each step.

Telegraphic Style

- The style of the key should be concise, similar to a telegraphic code. This brevity helps streamline the identification process, making it efficient and user-friendly.

Punctuation and Formatting

- Phrases in the key should be separated by semicolons, maintaining a logical flow throughout the key.

Diagnostic Characters in Couplets

- In the preparation of the key, the primary contrasting characters of each couplet should be diagnostic and definitive. These characters are the most reliable indicators of differences between taxa.

Supplemental Characters

- Alongside the primary diagnostic characters, it's important to include supporting supplemental characters that can provide further confirmation or clarification during the identification process.

Types of Taxonomic Keys

- Taxonomic keys are essential tools in the field of biology and taxonomy for the systematic identification of organisms. There are two primary types of taxonomic keys commonly used:

Indented Key

- The indented key is structured in a **hierarchical, indented format, with each choice leading to a more specific taxonomic category**. This format offers the advantage of quickly revealing the relationships between various taxonomic divisions. However, it can result in **lengthy keys with widely separated alternatives**, which may be considered inefficient in terms of space usage.

Bracket Key

- The bracket key, on the other hand, uses brackets to group related choices together. This format is particularly useful when dealing with **complex and intricately related taxonomic groups.It offers the advantage of presenting alternatives side by side, facilitating easy comparisons**. Moreover, the bracket key is often more economical in terms of space utilization as it does not employ indentation. When properly constructed, this key can be navigated forward or backward with equal ease, using numbers to indicate the pathway each choice follows. **However, a potential drawback is that the relationships between taxonomic divisions may not be immediately apparent upon visual inspection**.

- These two types of taxonomic keys each offer distinct organizational approaches, serving different needs and preferences in the realm of taxonomy. The choice of key type depends on the specific requirements of the taxonomic analysis and the complexity of the organisms being considered.
- Generic level identification keys are indispensable tools for taxonomists and researchers, offering a systematic and reliable method for determining the generic identity of organisms, making them an integral part of the taxonomic process.

UNIT – II

Chapter-4

Morphological, Morphometric and Meristic Characteristics of Finfish

Morphological Characteristics

- The external morphological characters of finfish are used for identification of species. There are two main features-morphometric characters and meristic characters. **Morphometric characters** include body shape/parts, mouth location and size, tail shape and colour. **Meristic characters** are the counting of spine, ray, etc in fin or other part of body.
- **External part of finfish** is commonly divided in **two parts**, 1) **Head** consist of Snout, Lips, Mouth, Jaws, Teeth, Barbels, Nostril, Eyes, Operculum and Gills; 2) **Body** incudes Fins, Lateral Line, Skin and Scales. present in these organs. These are detailed below.

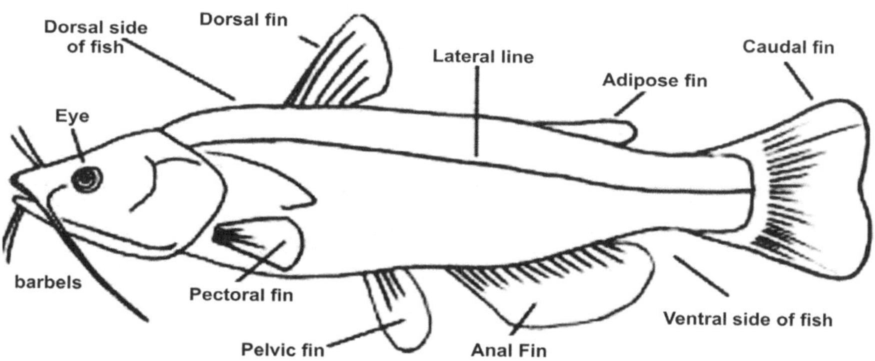

Finfish External Features

- **Section of finfish**: Finfish may be dissected into dorsal, ventral, anterior, and posterior ends.
- **Snout:** - Snout is the anterior most part of the fish (forward end of head). It is rounded or obtuse in most cases. There can be many variations to the shape of snout. 1. **Pointed** and sharp (Eels) 2. **Tubular** with jaws at tip 3. **Smooth** in most cases.
- **Lips**: -The bone of the upper and lower jaw is covered by lips. Mostly they are thin smooth membranes but, in some cases, they may have pores, stripes or modified to form a sucker like disc in Garra species.
- **Mouth**: - Mouth is the main organ which fish use while feeding. The position and shape of the mouth depend on the type of food a fish eats and the level at which it swims. The mouth shapes can be,

Terminal Mouth: - Terminal mouth is found in those fish, which feed from water column on other fish or zooplankton. Terminal mouths are located in the middle of the head and point straight forward; both jaws are of the same length. Example Danios, Rasbora, Puntius, etc.

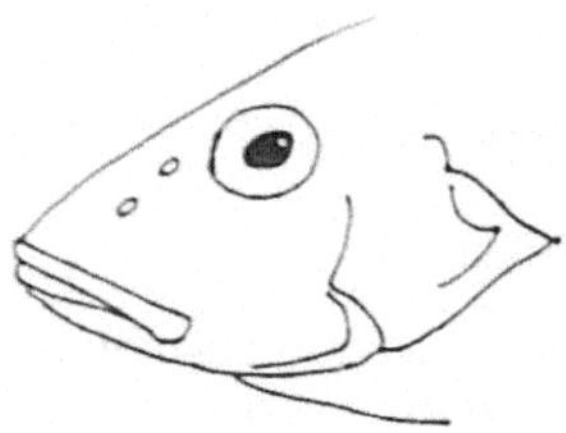

Superior Mouth: - The superior mouth is oriented upwards and the lower jaw is longer than the upper jaw. Usually, fish with this type of mouth feed at the surface. They lay in wait for prey to appear above them, then strike suddenly from below. Many species of fish with a superior mouth feed largely on insects, however, some may feed on other fish that swim near the surface.

Sub-terminal Mouth: - It is also called an inferior or ventral mouth. The inferior mouth is turned downward. The lower jaw is shorter than the upper jaw. Fish with inferior mouths are bottom feeders and often possess barbels that assist in locating food particles.

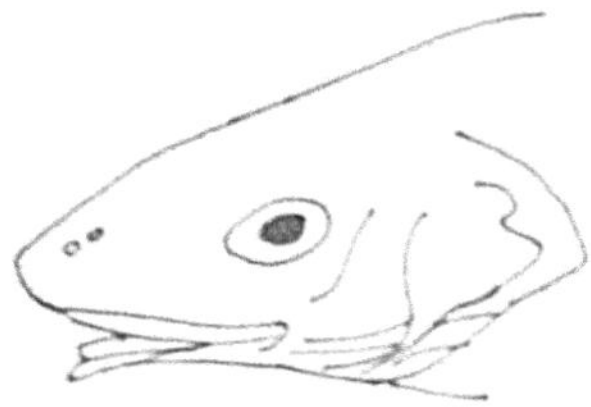

Protrusible Mouth: - Often fish will have a protrusible mouth feature, which allows them to extend their reach when attempting to snatch prey or food particles.

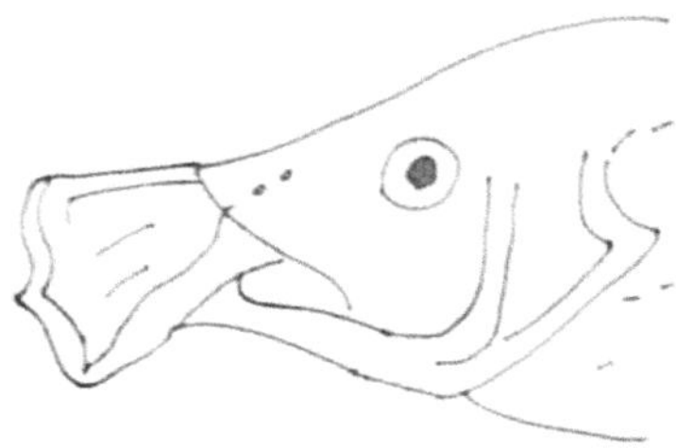

- **Jaws** :-The jaw consists of the upper jaw and lower jaw. The upper jaw consists of bones called pre-maxillaries and maxillaries, while the lower jaw consists of mandible bones. These are connected by a joint which enables the fish to open and close the mouth. Jaws contain teeth and frame the shape of the mouth.
- **Teeth**:- Most fish have teeth on jaws and palate. In addition to these teeth some fish have pharyngeal teeth also. However not all fish have teeth like cyprinids. There are many types of teeth :-
- **Canine: -** It is large conical teeth frequently located at the corners of the mouth, for example, snappers.
- **Viliform: -** Villiform teeth is small and fine teeth.

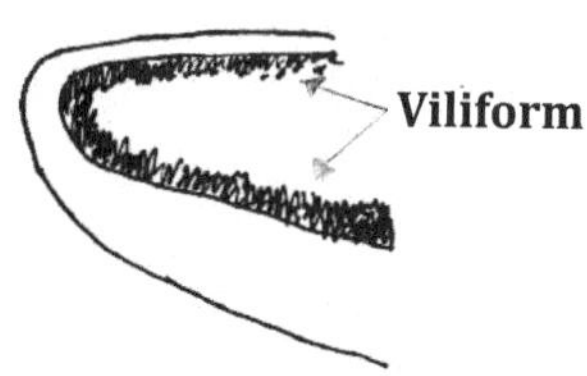

- **Molar form:** - Molariform teeth is molar like broad and rounded. It is used for crushing molluscs and crustaceans.

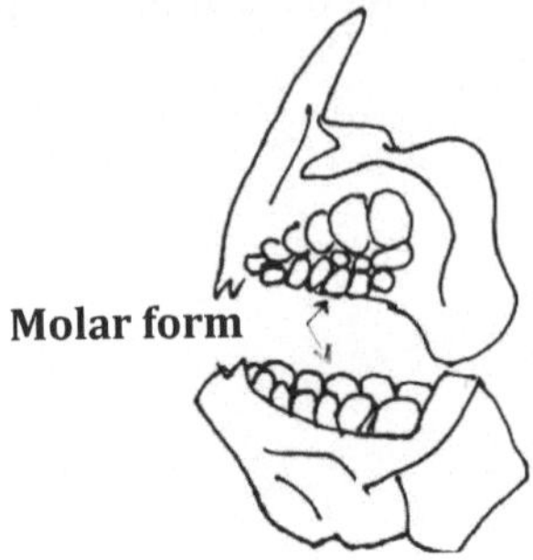

- **Cardiform:** - It is fine, pointed teeth arranged as in a wool card; for example, the pharyngeal teeth in pickerels (Esox).

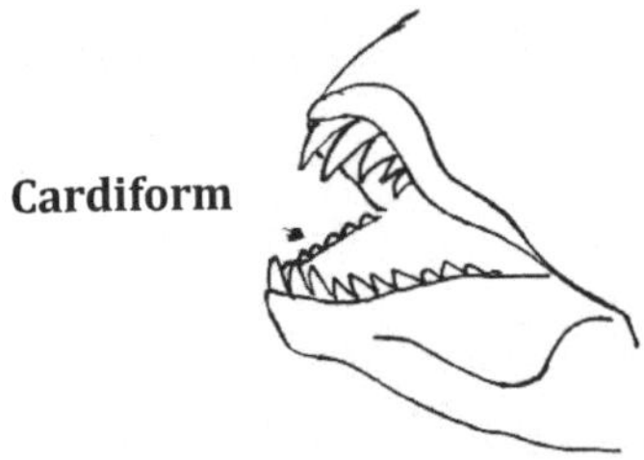

- **Incisor:** - Large teeth with flattened cutting surfaces adapted for feeding on molluscs and crustaceans; for example, chimaeras (Holocephali).
- **Barbels**: - Barbels are slender, whisker like tactile organs near the mouth. They are found in fish like catfish, carps etc, they house the taste buds used by fish to find food.

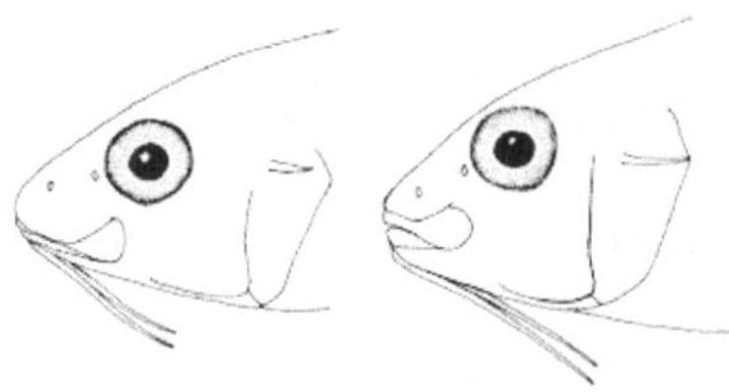

- **Nostrils**: - Nostrils are pair of apertures or slits on the snout of the fish. They are openings for the smell organs leading to the nasal canal on the skull. They are small to medium and are sunk in snout, in some fish like catfish they are covered with mucus. Position and type of nostrils is often used for identifying a fish.
- **Eyes:-** Eyes are mainly used by fish for seeing, food, enemies and predators. They are placed dorso-laterally (upper part- on the side) in most fish. However, the placement depends upon the habitat of fish. They can be placed at the top or bottom of the skull. Deepwater swimming fish have

large eyes as they receive less light at that depth, on the other hand eyes of hill stream fish are small as they live near the surface and receive a large amount of light.

- **Operculum and gills**:- Operculum along with gills form breathing apparatus for the fish. On each side of fish there are slits called gills. The gills are composed of comb-like filaments, the gill lamellae, which help increase their surface area for oxygen exchange. In bony fish, the gills lie in a branchial chamber covered by a bony operculum. A fish breathes by taking in water through its mouth and forcing it out from gills. There are tissue linings in the gills which absorb oxygen, CO_2 is also expelled from the gills. **The majority of bony fishes have 5 pairs of gills**. In cartilaginous fish, gill slits are not covered and lie in a row behind the head**. In general, there are five pairs in cartilaginous fish, but a few species have 6 or 7 pairs.**
- **Fins:-** Fins are thin appendages on the body of fish. They are made of bony spines protruding from the body of fish with skin covering them and joining them together in case of bony fish or Osteichthyes. In case of cartilaginous fish, fins are present as flippers. They are foil shaped and are primary means of locomotion for the fish. Some generate thrust when moved, others are used for stabilizing and treeing. In bony fish (Osteichthyes), most fins may have spines or rays. A fin may contain only spiny rays, only soft rays, or a combination of both. If both are present, the spiny rays are always anterior. Spines are generally stiff and sharp. Rays are generally soft, flexible, segmented, and may be branched. This segmentation of rays is the main difference that separates them from spines; spines may be flexible in certain species, but they will never be segmented. Fins occur in both pairs and single, they may be covered with scales or without scales.
- **Pectoral Fins:-** The pectoral fins occur in pair and are located on each side, usually just behind the operculum (gill cover), and are homologous to the forelimbs of tetrapods. There are many adaptions to these fins in some cases they create a dynamic lifting force that assists some fish, such as sharks, in maintaining depth and also enables the "flight" for flying fish. In many fish, the pectoral fins aid in walking, especially in the lobelike fins of some anglerfish and in the mudskipper.
- **Pelvic Fins:-** Pelvic fins occur in pairs and are found on the ventral (lower) side of the fish below the pectoral fins. They assist the fish in going up

and down in water, turning and stopping. In some fish like gobies they are joined into a single sucker like disc which is used to attach to objects.

- **Dorsal Fin**:- Dorsal fin is a single fin present on the dorsal side of body. The dorsal fin serves to protect the fish against rolling, and assists in sudden turns and stops. When the top of rays is connected with membrane they are called soft, else they are called hard spines. In many fish it is single and concave in shape with first spine being largest and last spine shortest. In perches there are two dorsal fins, one after another. The first fin is separated by either a short or long gap, or it may be combined.
- **Caudal Fin**:- Caudal or tail fin is located at the end caudal peduncle of the fish. The caudal peduncle is the narrow part of the fish's body to which the caudal or tail fin is attached. It is always a single fin and acts as rudder for the fish. On the basis of external and internal structure, caudal fins are Protocercal, Heterocercal, Diphycercal, Hypocercal, Homocercal, Isocercal and Gephyrocercal. The posterior end of the vertebral column is bent upward and continues upto the end of the caudal fin in the case of Heterocercal Caudal fin. Heterocercal caudal fin is found in elasmobranch (shark) and primitive fish such as Acipenser. Homocercal caudal fin looks externally symmetrical with equal epi-and hypochordal lobes, but internally hind end of the vertebral column is tuned upward and shortened. Thus, Homocercal caudal fin is internally asymmetrical. It is found in higher teleost.

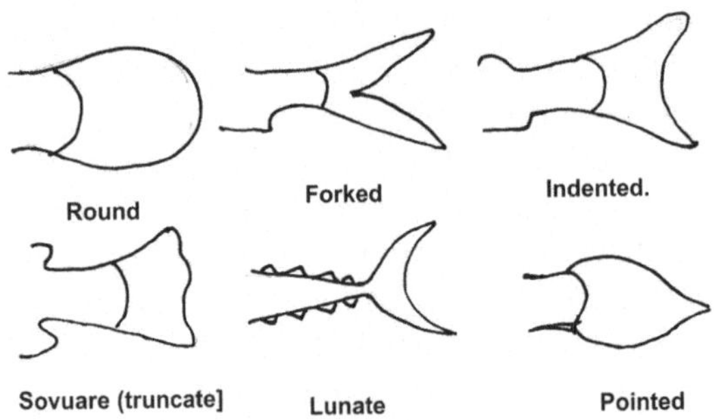

Types of Caudal fin

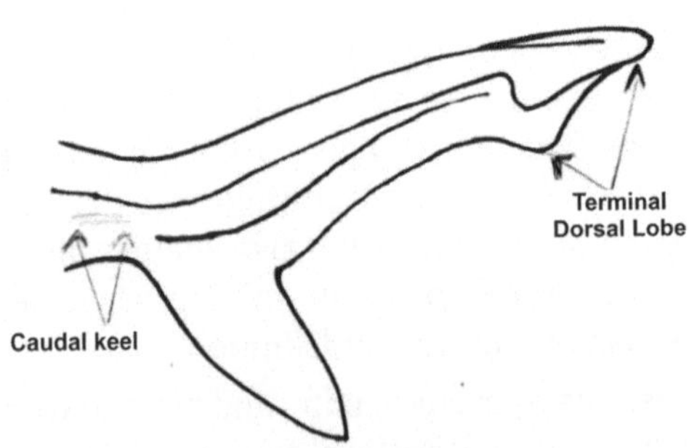

Heterocercal caudal fin

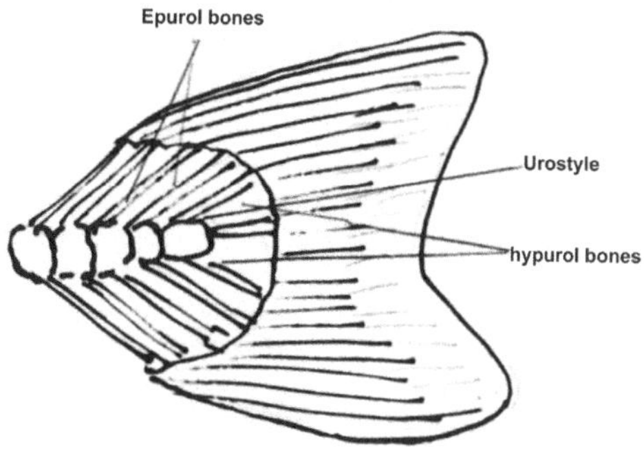

Homocercal caudal fin

- **Adipose Fins:-** The adipose fin is a soft, fleshy fin found on the back behind the dorsal fin and just forward of the caudal fin. It is absent in many fish families, but is found in Salmonidae, characins and catfishes.
- **Caudal Keel:-** Many fast swimming fish have a horizontal keel just in front of tail fin. It is present as ridge in caudal peduncle. It provides stability and support to the tail fin. They are always present as either a single pair or double pair, one of each side.
- **Fin lets:-** Fin lets are small fins, generally behind the dorsal and anal fins. Example, Tuna.
- **Skin:-**The skin of fish consists of live cells, there is very superficial amount of keratin in outermost layer. It is generally permeable. Fish typically have numerous individual mucus-secreting skin cells that aid in insulation and protection, but may also have poison glands, photospheres, or cells that produce a more watery, serous fluid. The colour in skin is mostly provided by melanin, however often the skin is colourless. The colour is provided by largely due to chromatophores in the dermis, which, in addition to melanin, may contain guanine or carotenoid pigments.
- **Lateral Line**:- The lateral line is a sense organ used to detect movement and vibration in the surrounding water. In most species, it consists of a line of receptors running along each side of the fish. It can be complete (*eg.* Pomfrets, Sciaenids), incomplete (*eg.* Pomacentridae) and interrupted (*eg.* Serranids, Cichlids).
- **Scale:-** The outer body of many fish is covered with scales. Some fishes have a type of scale known as the scute, which is an external shield-like bony plate, or a modified, thickened scale that often is keeled or spiny. eg. clupeids and carangids. There are four principal types of fish scales, which are following:
- **Placoid Scales**:- Placoid scales also called dermal denticles, are similar to teeth in that they are made of dentin covered by enamel. They are typically found in sharks and rays.

- **Ganoid scales**:- Ganoid scales are flat, basal-looking scales that cover a fish body with little overlapping. They are typical of gar and bichirs.
- **Cycloid scales:-** Cycloid scales are small oval-shaped scales with growth rings. They are found in many teleosts.
- **Ctenoid scales:-** Ctenoid scales are similar to the cycloid scales, with growth rings. They are distinguished by spines that cover one edge.

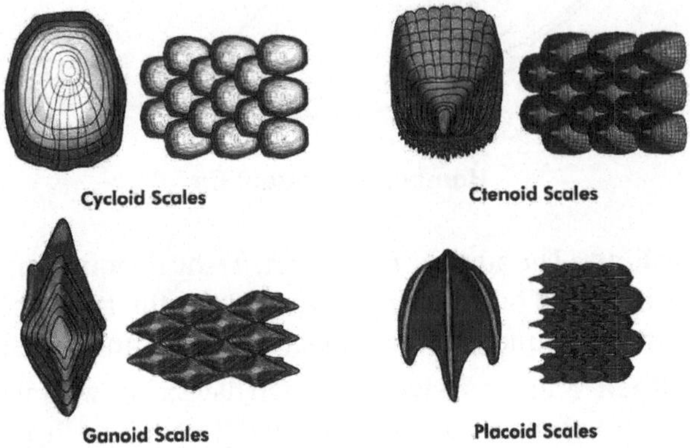

Body shapes

- **Anguilliform**: Greatly elongated snake like fish with an almost circular cross section eg. Eel
- **Taeniform** : Body is laterally compressed and greatly elongated bodies. eg. Gunnels
- **Ovate or Truncated**: eg. File fish and pomfrets
- **Depressiform**: Dorsoventrally compressed and bottom oriented fish. eg. skate, rays, toad fish.

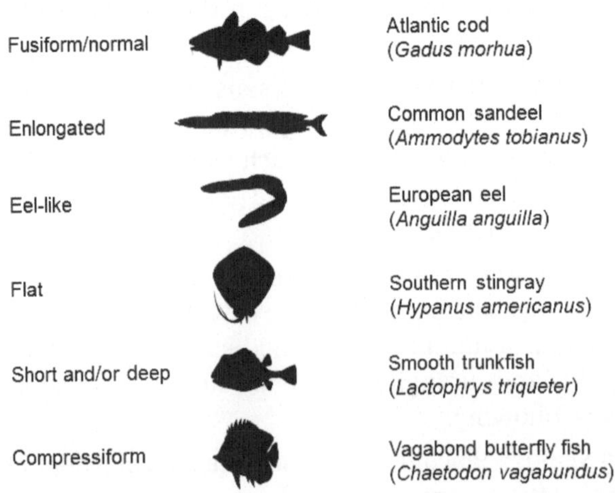

Morphometric Characteristics

- The morphometric characters are measurable features. These characters have been helpful for separating closely related genera and species and even population within species and are used in ichthyotaxonomical studies.

Measuring Fish Length

- Fish length measurements are typically taken with the fish positioned on its right side, snout to the left, using a measuring board. Here are the key methods and considerations.

Positioning on a Measuring Board

- The fish is placed on a measuring board equipped with a central scale and a headpiece against which the snout is gently pressed.
- It is essential to ensure the fish's mouth is closed, and its body and tail are aligned along the mid-line.
- Measurements are taken directly from the scale.

Precision and Tools:

- Measurements should be recorded to the nearest 0.5 mm.
- Fine draftsman dividers are commonly used to ensure accuracy.

Measuring Special Cases

- Rays and dorsoventrally flattened fish can be measured by lying straight on their ventral surface.
- In some cases, disc width is used as a linear dimension for rays.

Measuring Larger Fish

- For larger fish, calipers or a tape measure can be used for length measurements.
- It's crucial to measure from point to point along the body surface for accuracy.

Measuring Board Specifications

- To measure in centimeters, a 1 m board is sufficient, while a 30 cm extension piece can be added for larger specimens.
- A 50 cm board is usually suitable for measuring in half-centimeter units.

Scale Correspondence

- Ensure that the scale on the measuring board corresponds to the units being recorded.
- Avoid using boards with too many divisions, as this can lead to mistakes and unnecessary time consumption during recording.

Morphometric Measurements

- **Total Length (TOL):** The greatest dimension between the most anteriorly projecting part of the head and the farthest tip of the caudal fin when the caudal rays are spread out together.
- **Standard Length (STL):** The distance from the anterior most part of the head backward to the end of the vertebral column (structural base of caudal rays).
- **Fork Length (FOL):** Distance from the tip of snout to the end of the middle ray of the caudal fork when the fish is being flattened out.
- **Head Length (HEL):** Taken from the tip of the snout to the posterior most point reached by the bony margin of the operculum.
- **Pre-orbital length (PRO):** Distance from the tip of the snout to the forward point of eye.
- **Eye diameter (EYD):** Horizontal diameter of the visible part of the eye, i.e., the distance between the front edge and the back edge of the orbit.
- **Postorbital length (PSO):** Distance from the backward point of eye to middle of the backward bony edge of the operculum.
- **Upper jaw length (UPJ):** Length of maxillary is taken from the anterior most point of the premaxillary to the posterior point of the maxilla.
- **Lower jaw length (LOJ):** Length of lower jaw from anterior tip to angle of mouth.
- **Body depth (BDD):** Distance between the middle point of dorsal finbase to straight downward central margin of the body, excluding fins.
- **Pre-dorsal length 1 (PD1):** Distance from the tip of the snout to the forward origin of the dorsal (intersection point of the forward edge of the first ray of the dorsal, D1, with the outline of the back, the fish being flattened out)
- **Pre-dorsal length (PD2):** Distance from the tip of snout to the forward origin of the dorsal (intersection point of the forward edge of the first ray of the dorsal, D 2 , with the outline of the back, the fish being flattened out.
- **Pectoral fin length (PEL):** Distance from the extreme base of the uppermost ray to the farthest tip of the fin, filament if any.
- **Pelvic fin length (PVL):** Distance from the extreme base of the uppermost ray to the farthest tip of the fin, filament if any.
- **Dorsal fin length 1 (DF1):** Distance from the origin of the tip of the fin to the anterior lobe.
- **Dorsal fin length 2 (DF2):** Distance from the origin of the tip of the fin to the anterior lobe.

- **Inter dorsal length (IDL):** Distance from the base of the last spine (ray) of first dorsal to the intersection point of second dorsal fin.
- **Pectoral fin base length (PEB)**: Distance from the base of the anterior fin ray of the pectoral (P) to the backward end of the last ray, the pectoral being extended on the side of the fish in its normal position.
- **Pelvic fin base length (PVB):** Distance from the base of the anterior fin ray of the pelvic fin (P) to the backward end of the last ray, the ray being extended on the side of the fish in its normal position.
- **Dorsal fin base length (DB1):** Distance from the forward origin of the dorsal (D1) to the backward edge (Intersection point of the backward edge of the last spine, D', with the outline of the back, the fin being extended).
- **Dorsal fin base length (DB2):** Distance from the forward origin of the dorsal (D2) to the backward edge (Intersection point of the backward edge of the last ray, D2, with the outline of the back, the fin being extended.
- **Anal fin length (AFL):** Distance from the origin of the tip of the fin to the anterior most outer tip of the anal fin.
- **Anal fin base length (ABL):** Distance from the forward origin of the anal (A) to its backward edge (intersection point of the backward edge of the last ray, A' with outline of the abdomen, the fin being extended).
- **Caudal peduncle length (CPL):** Distance from the base of the second dorsal end to origin of the caudal fin.
- **Caudal peduncle depth (CPD):** Depth of the caudal peduncle.
- **Pre-pelvic length (PRP):** Distance from the tip of the snout to the anterior origin of the pelvic (intersection point of the forward edge of the first ray of the pelvic, with the contour of the abdomen, the fin being extended).
- **Pre-pectoral length (PRV):** Distance from the tip of the snout to the margin of the insertion of pectoral fin.
- **Pre-anal distance (PRA):** Distance from the tip of the snout to the forward origin of the anal (interior point of the forward edge of the first ray of the anal, A, with the outline of the abdomen, the fin being extended).

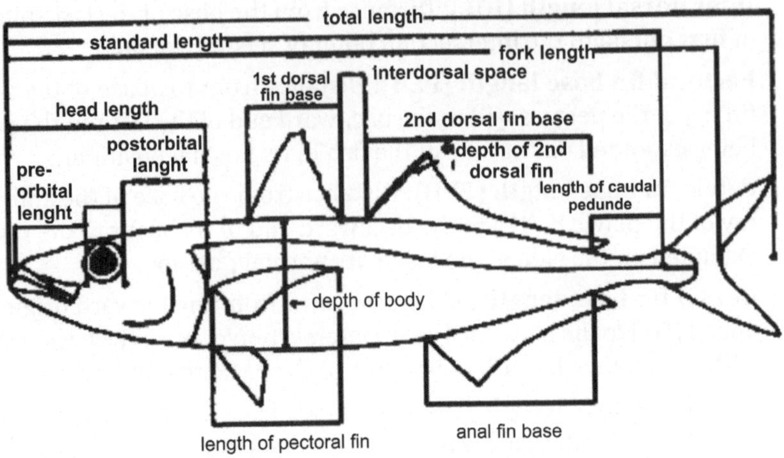

General Nomenclature of the External Morphology

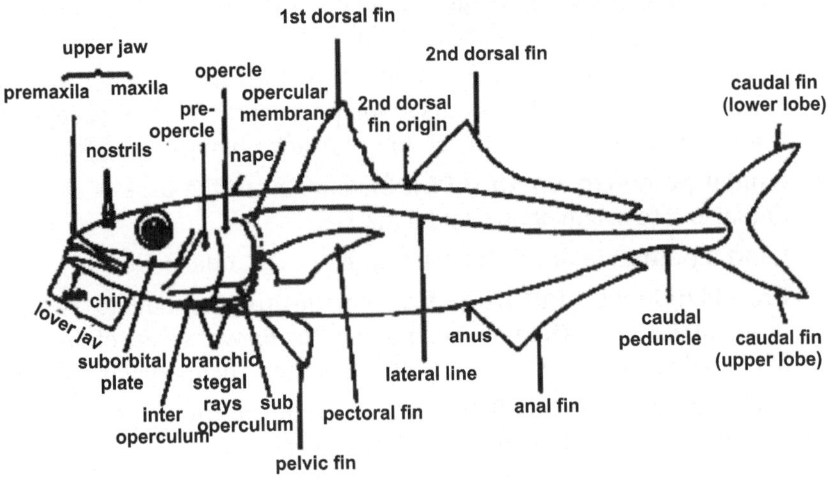

Meristics in Fish Identification

- Meristic characters are essential in the study of fish populations and species. Unlike characteristics such as body proportions or coloration, meristic characters are countable features that remain constant throughout an individual's life, typically established at or before metamorphosis. Here are key considerations in understanding meristics:

Counting Meristic Characters

- Meristic characters should be treated separately and include counts of features like fins, scales, and gill rakers.
- Frequency distribution of meristic characters should be provided to identify variations between species or populations of a species.

Abbreviations for Meristics in Teleost Fish

- **D:** Dorsal fin
- **A:** Anal fin
- **P1 or PF:** Pectoral fin
- **P2 or VF:** Ventral fin
- **C:** Caudal fin
- **Ll:** Lateral line scales
- **Ltr:** Lateral transverse row of scales
- **O:** Adipose dorsal fin
- **Gr:** Gill rakers

Dorsal Fin and Anal Fin Counts

- Both dorsal fin count and anal fin count include spines and rays.
- If there are two dorsals, one spinous and the other ray type, the formula is D1 and DII, where DI represents the first spinous dorsal and DII represents the rays of the second dorsal fin.
- For example, if there are 3 spines and 7 branched rays in a single dorsal fin, the formula would be DIII, 7.

Pectoral Fin and Pelvic Fin Counts

- Pectoral fin counts are typically made on the left side of the fish.
- In some cases, counts may be made on both sides to assess bilateral variations.
- Pelvic fin count includes both spines and rays if they are present.

Fin Count Formula Examples

- D1, I, VII-VIII: Denotes the first dorsal fin with one spine separated from the rest of the spines (VII-VIII).
- D2, I, 15-16: Denotes the second dorsal fin with one spine followed by 15-16 rays.
- AII, I, 10-15: Denotes the anal fin with two spines separated from one spine followed by 10-15 rays.

Gill Raker Counts

- Gill raker counts pertain to the lateral gill rakers located on the first arch, typically on the left side.
- The raker at the junction of the upper and lower limbs (epibranchial and ceratobranchial) is included in the lower limb count, as the major part of the base of the raker lies over the ceratobranchial.

- Rudimentary gill rakers with lateral base width equal to or less than the raker length, found at the anterior ends of the upper and lower limbs, are included in the counts but differentiated with specific notation, such as "ii, 7+19, iv=32."

Lateral Line Scales (Ll)

- Lateral line scales are scales along the lateral line, extending from its origin to the posterior-most part of the lateral line.
- It's important to note that in some teleostean fishes, like clupeids, the lateral line may be absent. In such cases, scales are counted along the row where the lateral line would typically be found.

Predorsal Scales

- Predorsal scales are scales on the midline in front of the dorsal fin origin.
- These scales are counted as the scale rows that intersect the midline from the anterior point of the dorsal fin to the orbit.

Scales Above and Below the Lateral Line (Ltr)

- This category refers to a transverse series of scale rows below and above the lateral line.
- Below the lateral line, scales are counted from the origin of the anal fin, excluding the median ventral scale row, along a forward diagonal to the lateral line.
- Above the lateral line, scales are counted from the origin of the dorsal fin, excluding the median dorsal scale row, on a diagonal backward to the lateral line.
- The lateral line row itself is not included in these counts.

UNIT – III

Chapter-5

Characters and Classifications of Finfish

Introduction

- Fish are a diverse group of vertebrate animals that have been around for over 400 million years, with over 40,000 species found in various aquatic environments. They come in different shapes, sizes, and have adapted to a wide range of habitats. The study of fish is known as "ichthyology," and it encompasses various aspects, including fisheries, aquaculture, and conservation.

Defining Fish

- Fish, in the context of this discussion, are aquatic animals with a backbone, gills for breathing, and fins for movement. While the term "fish" is often used loosely, it refers to these specific characteristics.

Economic Significance

- Fishes are economically significant as they provide food, liver oil, and various by-products such as fish meal. This makes fish biology an important field of study.

Common Characteristics of Fish

- Fishes exhibit the following common characters, most of which are related to their permanently aquatic mode of life.

- **Cold-blooded Vertebrates**: Fish are cold-blooded and often have a streamlined body, although some have snake-like or flattened shapes.
- **Respiratory Organs:** They breathe through gills.
- **Fins for Locomotion:** Fish have paired and unpaired fins, including pectoral, pelvic, and dorsal and ventral median fins. The tail has a caudal fin for propulsion.
- **Body Covering:** Their bodies are typically covered with scales, but some are scaleless.
- **Skeletal Structure:** They have a backbone made up of either cartilaginous or bony vertebrae, replacing the notochord.
- **Visceral Arches:** Fish have paired visceral arches. The first pair forms their jaws, the second forms the suspensorium, and the remaining five pairs support the gills.
- **Endoskeleton:** The internal skeleton can be either cartilaginous or bony.
- **Middle Ear:** Fish lack a middle ear.
- **Kidneys:** They have mesonephric kidneys in adulthood.
- **Lateral Line Sense Organ**: Fish possess well-developed lateral line sense organs, which help them detect vibrations and movement in the water.
- **Heart:** Their hearts typically have one auricle and one ventricle and function as venous pumps, sending blood to the gills for oxygenation.
- **Cranial Nerves:** There are ten pairs of cranial nerves in fish.

These common characteristics are adaptations that have allowed fish to thrive in a variety of aquatic environments and make them a fascinating subject of study in the field of biology.

Classification of Fish

The classification of fish has evolved over time as our understanding of these creatures has expanded. Different taxonomists have proposed various schemes, but there are some key milestones in the development of fish classification.

Early Classification Based on Morphology

Early taxonomists primarily relied on the physical characteristics (morphological characters) of living fish species for their classification. This approach was limited to studying the external features of fish.

Incorporating Fossil Forms and Cladistics

As scientific knowledge advanced, later taxonomists began considering not only living fish species but also their fossil relatives. They also started using cladistic principles, which analyze the evolutionary relationships among organisms based on shared characteristics.

Contributions of Notable Taxonomists

Several prominent taxonomists have significantly influenced the classification of fish:

- **Berg (1940) and Romer (1959):** They provided detailed classifications of fish, considering both living and fossil forms. Their classifications have been widely accepted in the scientific community.
- **Bertin and Arambourg (1958) and Greenwood et al. (1966):** These researchers introduced important modifications to existing classification schemes, adding to our understanding of fish diversity.

The Modern Classification of Fish

In more recent times, several taxonomists have proposed a modern classification of fish, which represents the current state of understanding. Notable contributors to this modern classification include:

- Lauder and Liem (1983)
- Nelson (1984)
- Pough et al. (1989)

This modern classification incorporates the latest scientific knowledge and methods, reflecting the evolutionary relationships and diversity of fish species. It's considered a comprehensive and up-to-date framework for understanding the classification of fish.

Historical Changes in Fish Classification

The classification of fish has seen significant changes over time, with different taxonomists proposing various schemes based on evolving knowledge. Here's a breakdown of these developments:

Berg (1940)

- Berg's classification considered both recent and fossil forms of gnathostomes (jawed vertebrates) with gills and fins for locomotion.
- **He divided them into seven classes:** Acanthodii, Coccostei, Pterychthyes, Elasmobranchii, Holocephali, Dipnoi, and Teleostomi. The first three groups were completely extinct, known as Placoderms.
- Actinopterygians (ray-finned fishes) were originally divided into three groups: Chondrostei, Holostei, and Teleostei, based on morphological differences.

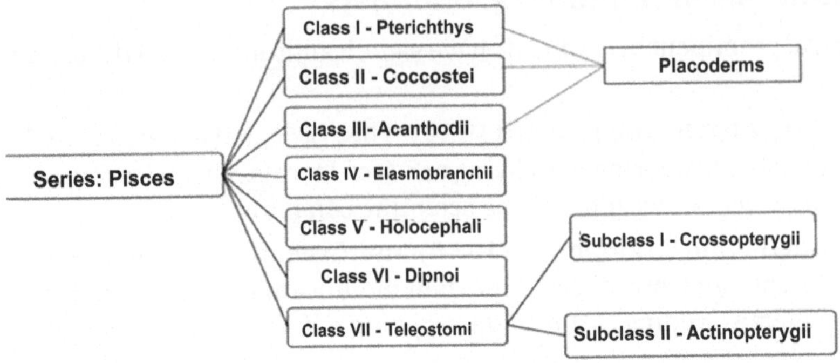

- Berg's scheme abolished these divisions and replaced them with a series of orders.

Romer (1959)

- Romer introduced a different classification, combining Elasmobranchii and Holocephali into one class, Chondrichthyes (cartilaginous fishes).
- He grouped all bony fishes into one class, Osteichthyes, which included Dipnoi and Teleostomi from Berg's classification.

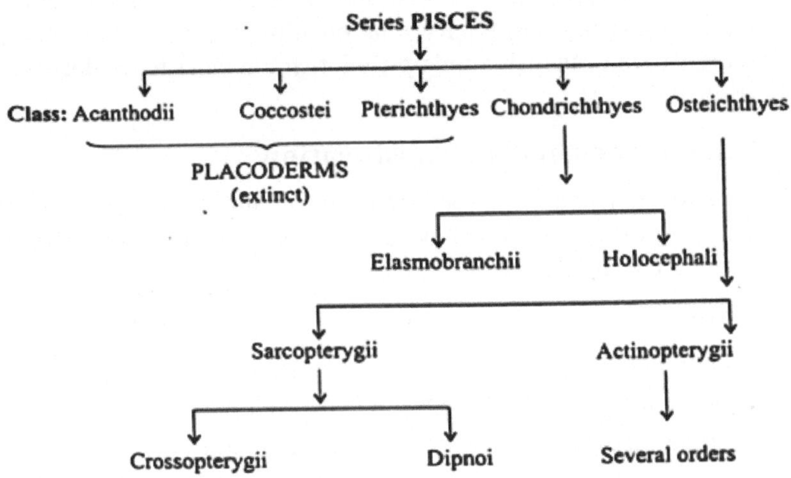

- Romer divided Osteichthyes into two sub-classes: Sarcopterygii (including Crossopterygii and Dipnoi) and Actinopterygii (all bony fishes without lobate fins).

Bertin and Arambourg (1958)

- They disagreed with Berg's classification of Crossopterygii and Dipnoi as orders within the sub-class Sarcopterygii, considering it an error.
- They proposed that Crossopterygii and Dipnoi form independent sub-classes along with Actinopterygii.

- They also suggested a new sub-class called "Brachiopterygii" for fishes like Polypterus and Chalamoichthys, which differ significantly from other actinopterygians.

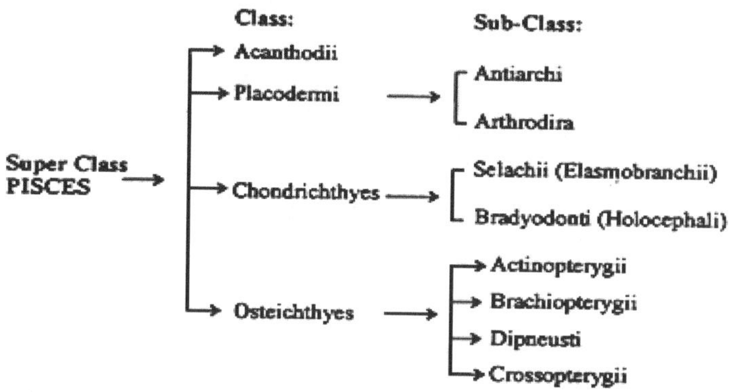

Modern Classifications

Greenwood et al. (1966) introduced a division of teleosts into three main divisions with eight superorders, but their scheme has seen refinements.

Such as Super orders

- **Division I-** Eleomorpha and Clupeomorpha
- **Division II-** Osteoglossomorpha
- **Division III-** Paracanthopterygii, Ostariophysi, Protocanthopterygii, Atherinomorpha and Acangthopterygii.

Lauder and Liem (1983), Nelson (1984), and Pough et al. (1989) divided teleostean fishes into four groups known as **"Infra-divisions."**

- **Lauder and Liem (1983)** Divided Infra Divisions such as 1. Osteoglossomorpha 2. Eleomorpha 3. Clupeomorpha 4. Euteleostei (Paracanthopterygii, Ostariophysi, Protocanthopterygii, Atherinomorpha and Acangthopterygii.)
- **Pough et al. (1989)** his classification similar like Lauder and Liem but he added extra Scopolomorpha Super order into Euteleostei category.

Conclusion

- The classification of fish has evolved over time. Early classifications were based on the morphology of living species. Later, classifications incorporated fossil forms and cladistics. Key taxonomists like Berg and Romer proposed their own systems. Challenges arose in classifying teleostean fishes. Modern classifications, including those by Lauder, Nelson, and Pough, have refined the categorization, aiming to accurately represent the diversity of fish species. Classification continues to evolve with advancing knowledge.

Chapter-6

Ostracoderms and Placoderms

Ostracoderms

- Agnatha is a group of ancient vertebrates that appeared in the Ordovician period and thrived during the Silurian and early Devonian periods.
- They are notable for the absence of jaws and are often referred to as jawless fish. Most of the Agnatha are now extinct, with the exception of Cyclostomata, which includes modern lampreys and hagfish.
- The fossil Agnatha were heavily armored, and their bodies were protected by external bony plates. They are known as "ostracoderms," where "ostrac" means shell and "derm" means skin.
- Unlike more modern vertebrates, they did not have a bony vertebral column or spine. Instead, some species had a single pair of fins located behind the head.

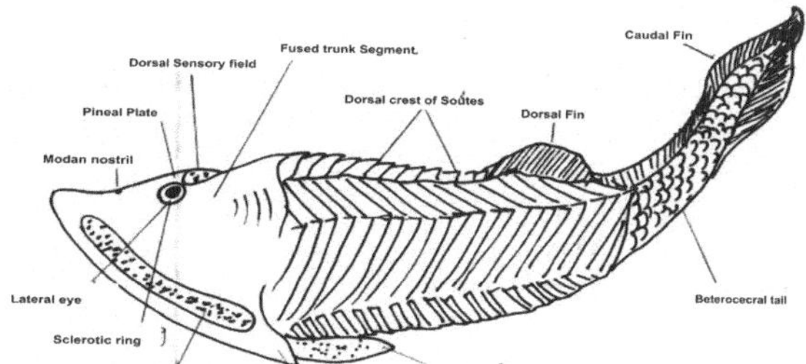

Agnatha can be Further Divided into Five Classes

- **Euphanerida:** An extinct class of Agnatha, with various species displaying unique features and adaptations.
- **Pteraspida:** Another extinct class of Agnatha, characterized by their distinct physical attributes and adaptations.
- **Anaspida:** This class is yet another group of extinct Agnatha, which had its own set of characteristics and evolutionary history.
- **Cephalaspida:** Like the other classes, Cephalaspida represents an extinct group of jawless fish-like vertebrates with their own specific characteristics.
- **Cyclostomata:** This is the class of Agnatha that has survived to the present day and includes modern lampreys and hagfish. These creatures have retained the jawless characteristic and provide insights into the evolutionary history of Agnatha.

Cyclostomes: The Most Primitive Living Craniates

Cyclostomes, which include lampreys and hagfishes, are the modified descendants of the Agnatha. They represent the most primitive living craniates and exhibit a low level of organization. Here are the key distinguishing features of Cyclostomes:

- **Body:** Their bodies are elongated, cylindrical, and covered with smooth, slimy skin. Scales are absent.
- **Jaws:** Cyclostomes lack jaws; instead, they have circular, suctorial mouths equipped with horny teeth and a rasping tongue.
- **Respiration**: They respire through pouch-like or sac-like gills.
- **Paired Fins**: Cyclostomes lack paired fins.
- **Sensory Organs**: They have a single nasal aperture and a median olfactory organ.
- **Skeleton:** The notochord is persistent, and their skeletons are cartilaginous. A branchial basket supports the pharynx.
- **Digestive System**: Their alimentary canal is straight, and they lack a cloaca.
- **Nervous System**: The brain has a small cerebellum, and the sympathetic nervous system is absent.
- **Hearing:** They have one or two semi-circular canals in the ear.

Cyclostomata, Which Means "round mouth," is Divided into two Sub-Classes

Petromyzontia (Lampreys)

- They have one or two well-developed dorsal fins and a complete branchial basket.
- Seven gill apertures are present on each side, opening into a respiratory tube below the oesophagus.

- ❑ The naso-hypophyseal cavity does not open behind into the pharynx.
- ❑ Lampreys can be found in both marine and freshwater environments.
- ❑ **Examples:** *Petromyzon marinus* (sea lamprey), *Lampetra fluviatilis* (freshwater lamprey).

Myxinoidea (Hagfishes)

- ❑ They are exclusively marine cyclostomes.
- ❑ Dorsal fins are feebly developed or completely absent.
- ❑ They have terminal mouths, and 1-15 gill apertures are present on each side, opening directly into the pharynx.
- ❑ The branchial basket is rudimentary, and the naso-hypophyseal cavity opens behind into the pharynx.
- ❑ **Examples:** *Myxine glutinosa, Bdellostoma (Eptatretus) cirrahatum.*
- ❑ Hagfishes are typically found along the coasts of Northern Europe, North and South America, New Zealand, and South Africa.

S.No	Lamprey	Hagfish
1.	Possses one or two dorsal fins	Dorsal fin absent
2.	A suctorial buccal funnel is present	No buccal funnel
3.	Eyes are large	Eyes are rudimentary
4.	Barbles are not present	3 pairs of barbles are present
5.	A single nasal aperture is present on the dorsal side of head	Nasal aperture lies infront of head
6.	Seven gill apertures on each side	1-14 pairs of gill openings
7.	Awell developed cartilaginous cranium is present	Poorly developed cranium
8.	Vertebrae are cartilaginous	Vetebrae absent
9.	Well-developed branchial basket	Rudimentary
10.	Nasohypophyseal sac does not open into the pharynx	It opens into the pharynx
11.	A pronephric kidney is absent in the adult	Present
12.	Lateral line organ is well developed	These are degenerate
13.	Eggs are small, without hooks metamorphosis take place.	Eggs are large with hooks. There is no metamorphosis.

Placoderms: The Palaeozoic Fishes

- ❑ Placoderms were among the earliest jawed vertebrates, characterized by their heavy external bony plates. They thrived during the Devonian period but went extinct by the end of the Permian period. Since they lived during the Palaeozoic era, they are often referred to as "Palaeozoic fishes."

Placoderms had specialized adaptations for various lifestyles and shared common characteristics

- ❑ They had primitive jaws with an aphetohyoidean condition, meaning the hyoid arch did not support the jaw.

- The upper jaw was fused with the skull.
- They possessed hyoid gill slits.
- Placoderms had paired fins.
- They lacked spiracles.

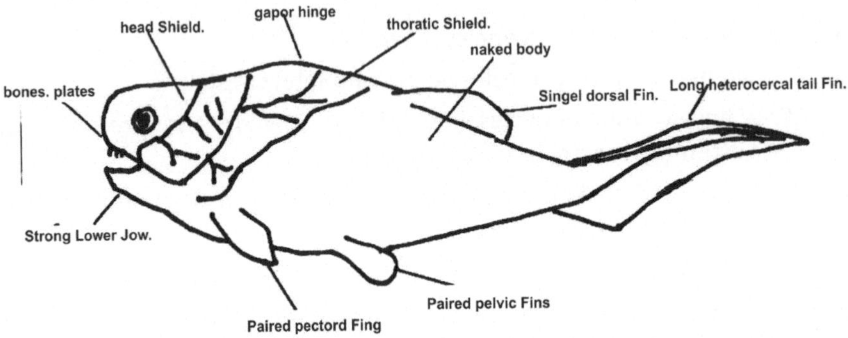

Placoderms existed for a relatively short time in the evolutionary history and are considered as "unsuccessful ancient experiments" in the evolution of jawed vertebrates. They can be divided into three main classes:

- **Acanthodii**- Also known as "*spiny sharks*," these were small-sized placoderms. A notable example is Climatius, which lived during the late Silurian and early Devonian period. Climatius had a fish-like appearance and was covered in small rhombic scales.
- **Coccostei (Arthrodira)** - These armored fishes were the most successful vertebrates of the Devonian period but eventually became extinct. Coccosteus, a well-known example, was a fast-swimming creature, measuring about 60 cm in length.
- **Subclass**

Macropetalichthyes - This subclass includes genera like Macropetalichthys and Lunaspis from the Devonian. They resembled arthrodires with a strong head shield followed by thoracic plates and paired pectoral spines, although they lacked a ball and socket joint between the head and body shield. Their paired fins were similar to those of elasmobranchs.

- **Pterichthyes (Antiarchi)** - This class is represented by Pterichthys and Bothriolepis. Like arthrodires, they had a bony shield for the head and thoracic region, but the arrangement of cephalic plates and the mode of attachment of the head with the body shield were different.

In addition, Class Comes Under Placoderms Mentioned Below

- **Stegoselachii**

Gemuendina- Shark-like Appearance: Gemuendina, a representative of Stegoselachii, bore a striking resemblance to modern sharks. It had a flat body with a broad head.

- **Palaeospondyli**- Palaeospondylus is the sole representative of this group. It was a small fossil from the mid-Devonian, around 5 cm long. Its affinities are not fully known, but it might be a small placoderm that lost its external bony armor. Some scientists think it could be related to Agnatha due to the presence of a cartilaginous roof, while others compare its jaw apparatus to Acanthodii, suggesting it was a true gnathostome.

S. No	Evolution
1	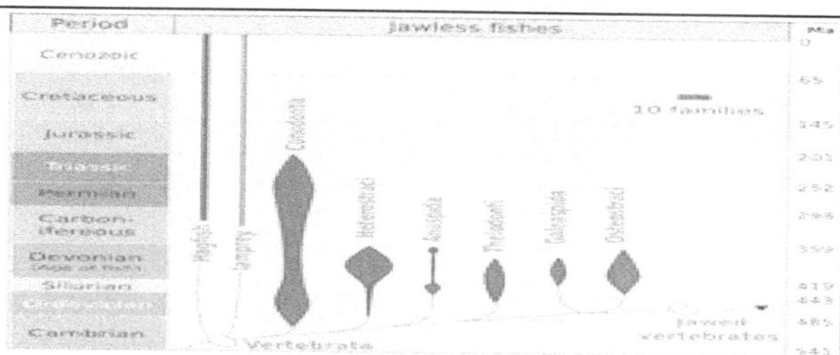
2.	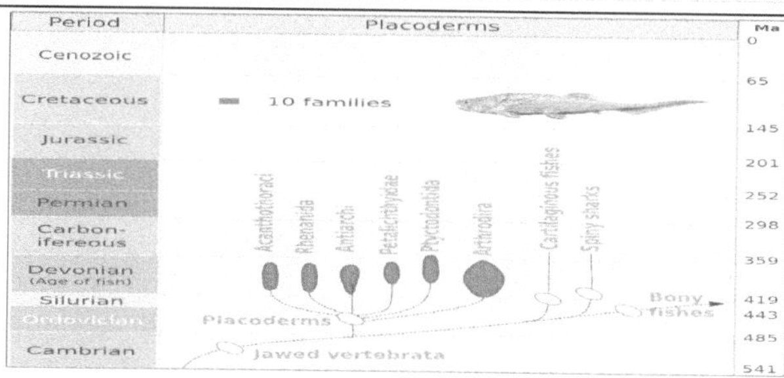
3.	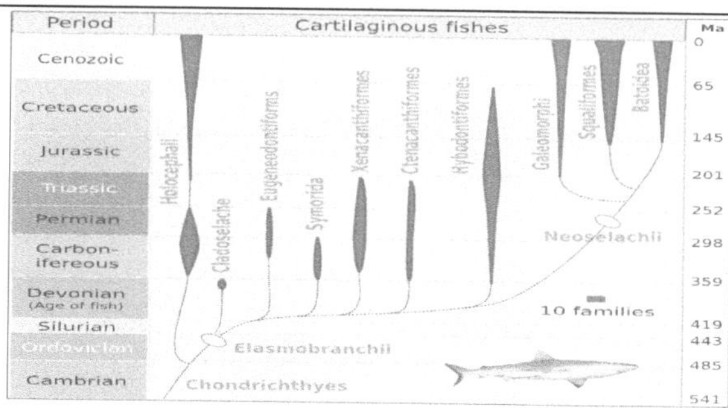
4.	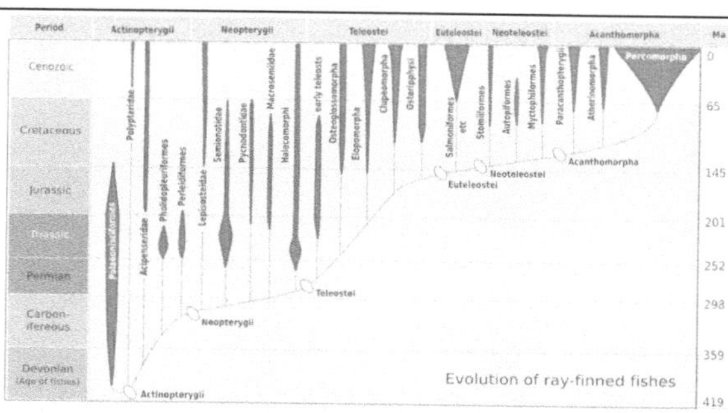

Chapter-7

Chondrichthyes

Chondrichthyes: The Cartilaginous Fishes

Introduction

Chondrichthyes, a prominent group of fishes, encompassing species such as sharks, skates, and rays, are characterized by a cartilaginous endoskeleton and are primarily found in marine environments. They exhibit several distinctive features:

- **Exoskeleton**: Their exoskeleton takes the form of dermal denticles or placoid scales.
- **Endoskeleton**: They lack membrane bones, possessing a cartilaginous endoskeleton.
- **Gill Slits**: Each side of their body features five to seven gill slits.
- **Operculum**: Unlike most fishes, they lack an operculum, so their gill slits open directly to the exterior.
- **Tail**: Chondrichthyes typically have a heterocercal tail.
- **Spiracle:** Many species have a spiracle.
- **Claspers**: Males have a pair of claspers used in reproduction.
- **Air Bladder:** They lack an air bladder.
- **Intestine:** Their intestine is characterized by a spiral valve.
- **Cloaca:** A cloaca is commonly present in these species.
- **Urea in Blood:** Their blood has a high concentration of urea.
- **Heart:** Their heart features a contractile conus with several valves arranged in transverse rows.

- **Venous System**: The venous system is enlarged into sinuses at various points.
- **Reproduction:** Fertilization is internal, and many Chondrichthyes species are viviparous.

Chondrichthyes are often predatory, with some growing to impressive sizes. For instance, Carcharodon can reach lengths of about 25 meters, while the Rhinodon (whale shark) reaches 20 meters, and the Cetorhinus (basking shark) can attain lengths of approximately 12 meters. These giants primarily feed on plankton and pose no threat to humans. However, species like the white shark *(Carcharodon carcharius)*, blue shark (Carcharinus), and tiger-shark (Stegostoma) are known for their predatory behavior.

In the fossil record, cartilaginous fishes, including sharks, rays, and ratfishes, first appear in the early Devonian period. They have undergone various morphological adaptations and specializations while retaining ancestral characteristics. They can be categorized into two groups:

Holocephali (Rat Fish and Chimaera): This group has a single external gill aperture.

The ratfish, also known as chimaera, is a type of cartilaginous fish belonging to the subclass Holocephali. Ratfishes have distinctive characteristics that set them apart from other cartilaginous fish like sharks and rays. These characteristics include:

Four gill openings: Ratfish have four gill openings, unlike some other cartilaginous fish, which may have more.

Spiracle absent: Unlike some sharks that possess a spiracle (an additional small respiratory opening), ratfish lack this feature.

Upper jaw fused to the skull: The upper jaw of ratfish is fused to the skull, a feature that differentiates them from sharks.

Few, large, flat teeth: Ratfish have relatively few teeth compared to many other cartilaginous fish, and their teeth are flat plates.

Scales absent: Ratfish lack the scales commonly found on the skin of other fish.

Claspers: Male ratfish possess claspers, which are specialized organs used in copulation.

Elasmobranchii (Plate-Gilled): Elasmobranchii can be further divided into two subgroups:

- **Pleurotremata (Sharks)**: These species have a cylindrical body with seven gill slits on each side of the head.
- **Hypotremata (Skates and Rays)**: This sub-group exhibits a dorsoventrally flattened body and features five pairs of gill slits on the ventral side.

Elasmobranch Fishes: A Brief Overview

Sharks, rays, and skates belong to the class Chondrichthyes and are collectively known as cartilaginous fishes.

Approximately 600 to 700 species of these fishes are found in waters worldwide. In contrast to their bony fish counterparts, cartilaginous fishes are often regarded as more primitive.

Some members of this group exhibit unique specializations, similar to the distinct adaptations seen in teleosts among bony fishes.

Extinct Shark-Like Fishes of the Paleozoic Era

During the Paleozoic era, two noteworthy genera of extinct shark-like fishes left their mark: Cladoselache and Xenacanthus.

Cladoselache: A Primitive Predator

Cladoselache, depicted in, reached lengths of up to 2 meters. It featured distinct characteristics:

- **Terminal Mouth**: Cladoselache had a terminal mouth.
- **Heterocercal Tail**: Its tail was heterocercal, with the upper lobe larger than the lower lobe.
- **Persistent Notochord**: Cladoselache retained a notochord, the flexible rod-like structure running the length of the body.
- **Five Gill Apertures:** It had five gill apertures for respiration.
- **Unique Pectoral Fin:** The large pectoral fin was structurally primitive, featuring a broad base without the typical anterior and posterior constrictions separating the fin from the body.
- **Dorsal Fin Spines:** Stout spines preceded the first, and sometimes the second, dorsal fins.
- **Pelvic Fins**: Although smaller, the pelvic fins had a structure akin to the pectoral fins.
- **Caudal Fin:** Externally symmetrical, the caudal fin had an asymmetrical internal structure.
- **Efficient Predatory Teeth:** Cladoselache possessed sharp, pointed teeth ideally suited for piscivorous feeding on other fishes.

Xenacanthus (Pleurocanthus): A Specialized Predator

Xenacanthus, also known as *Pleurocanthus*, exhibited a higher degree of specialization compared to Cladoselache. It featured the following characteristics.

- **Archipterygeal Pectoral Fin:**Xenacanthus had an archipterygeal pectoral fin, consisting of an elongated axis with pre and post-axial rays.
- **Dorsal Fin Arrangement**: It had a long dorsal fin along with two smaller fins.

- **Serrated Head Spine**: Xenacanthus sported a long, serrated spine on its head.
- **Diphycercal Tail**: Its tail was diphycercal, with a symmetrical upper and lower lobe.
- **Claspers in Males**: Male Xenacanthus featured claspers, specialized structures for reproduction.
- **Persistent Notochord**: Similar to Cladoselache, Xenacanthus retained a persistent notochord.
- **Absence of Scales**: Unlike many contemporary fish species, Xenacanthus lacked scales.

The characteristics that set cartilaginous fishes apart include:

- **Notochord Constriction**: Their notochord is constricted by vertebrae.
- **Cartilaginous Skeleton:** Elasmobranchs have a cartilaginous skeleton, with calcified cartilage that resembles bone.
- **Absence of Swim Bladder:** Unlike bony fishes, they lack a swim bladder.
- **Skull Structure:** In living forms, the skull lacks sutures.
- **Teeth:** Their teeth are usually not fused to the jaws and are replaced serially.
- **Nasal Openings**: Nasal openings are usually single and more or less ventral in position on each side.
- **Intestinal Spiral Valve:** They possess an intestinal spiral valve, aiding in digestion.
- **Fertilization:** Fertilization can be either external or internal, with males often having pelvic claspers.
- **Embryo Encapsulation:** Elasmobranch embryos are encapsulated in a leather-like case.

S.No	Characters	Subclass: Elasmobranchii	Subclass: Holocephali
1.	Important Info.	Elasmobranchii encompasses sharks, rays, and skates, with 128 genera and 608 living species.	Holocephali, or ratfishes, include species with a long slender tail, primarily represented by chimaeras.
2.	Gill Openings	Typically, elasmobranchs have five to seven gill openings, with the presence of a spiracle (though secondarily lost in some species).	Typically, four gill openings, with a spiracle absent.
3	Jaws	The upper jaw is not fused to the cranium but attached with either amphistylic or holostylic suspension.	The upper jaw is fused to the skull.
4.	Scales	Their bodies are covered with placoid scales.	Scales are absent.

S.No	Characters	Subclass: Elasmobranchii	Subclass: Holocephali
5.	Teeth	They have numerous teeth.	They have few, large, flat plates.
6.	Cloaca	A cloaca is present.	A cloaca is present.
7.	Mating Structure	Males often have intromittent organs	In males, claspers are seen on the head region, in addition to pelvic claspers.

Elasmobranchii is further divided into super-order , including:

S.No	Super-order	Characters
1.	Galeomorphi	Galeomorphi sharks exhibit diverse body shapes, including species that differ markedly from the typical shark body shape. Order: - Heterodontiformes: Known as horn sharks, these are shallow water bottom dwellers. Lamniformes: Includes various shark families, with many large pelagic predators.
2.	Squatinomorphii	Squatinomorphii is represented by angel sharks. These species exhibit features intermediate between sharks and rays.
3.	Squalomorphii	Squalomorphii includes shark species with a flattened body but distinct from rays. This superorder encompasses orders such as Hexanchiformes, Squaliformes, and Pristophoriformes, each with its unique characteristics.
4.	Batoidea	Batoidea comprises rays and skates, which are adapted for bottom living and benthic habitats. Rajiformes, Pristiformes, Torpediniformes, and Myliobatiformes: Each order within this superorder features rays and skates adapted for life on the ocean floor.

Sub-classes: Beyond Elasmobranchii

Sub-class Dipneusti: Lungfishes

This sub-class includes freshwater lungfishes with a long evolutionary history. Notable genera include Neoceratodus, Lepidosiren, and Protopterus.

Sub-class Crossopterygii: Fringe-Finned Fishes

The fringe-finned fishes are represented by the living species Latimeria chalumnae, known for its discovery in South Africa and the Comores Archipelago.

Sub-class Brachiopterygii: A Single Family

This sub-class is represented by the Polypteridae family, which includes 10 species. These fish have ganoid scales and possess a spiracle, and they use both gills and lungs for respiration.

Sub-class Actinopterygii: Ray-Finned Fishes

The Actinopterygii subclass includes the majority of bony fishes, divided into three infra classes: Chondrostei, Holostei, and Teleostei, each with its unique features.

Chapter-8

Sharks

Introduction

Shark Species Diversity

Sharks, belonging to the group called elasmobranchs, are ancient creatures with a history spanning over 400 million years. In Indian waters, there are 48 shark species, categorized into 6 contributing major fisheries, 19 minor fisheries, and 23 non-fishery species. The Order Carchariniformes encompasses 33 species, while the remaining 115 represent 4 taxonomic orders.

Shark Distribution

Primarily marine, sharks also enter bays, lagoons, estuaries, and some, like *Carcharhinus leucas* and *Glyphis gangeticus,* migrate into rivers and freshwater lakes. Their habitats range from pelagic and epipelagic to demersal, with depths reaching up to 1,000 meters.

Shark Life Patterns

Shark life is characterized by slow growth, late reproduction, and the birth of a few well-developed young ones. Viviparous species release limited but well-developed young, showcasing variations in size and reproductive patterns. Some species, like the bull shark, release young at lengths representing a significant percentage of the adult size.

Reproductive Strategies

Sharks exhibit varied reproductive strategies, with 34 species being viviparous, 8 ovoviviparous, and 6 oviparous. Some species, like the whale shark, still pose questions regarding their reproductive processes. Oviparity is considered primitive, while viviparity involves retaining internally fertilized ova. Sharks mature late, with first maturity occurring at 52.2 to 85.4% of their maximum length.

Economic Exploitation and Conservation

Sharks face target fishing due to the economic value of their various body parts. Meat, liver, fins, hide, jaws, and teeth are all utilized. Shark fins, especially prized, contribute significantly to exports. Concerns about overfishing led to protective measures, placing certain shark species under Schedule I of the Wildlife (Protection) Act, 1972, highlighting the need for conservation efforts.

Growth and Fishing Impact

Sharks exhibit slow growth rates compared to teleosts. Growth data is available for only a few species, indicating annual growth rates ranging from 0.33 to 1.08. The relationship between the stock and recruitment is crucial, considering low fecundity and the dependence on the number of adults in the stock. Over recent years, sharks have become targets for fishing due to the lucrative market for various body parts.

Economic Significance and Exports

Sharks are economically valuable for their meat, oil, fins, hide, jaws, and teeth. Shark fins, especially sought after in oriental cuisine, are among the most expensive fish products globally. India exported significant quantities of dried shark fins, frozen shark meat, squalene, shark skin, and dried shark bones, contributing to the country's economy.

Conservation Measures

To protect elasmobranchs, the Indian government categorized four shark species, including the whale shark and Pondicherry shark, under Schedule I of the Wildlife (Protection) Act, 1972. This conservation effort extends to rays, sawfishes, and guitarfishes. Concerns about the impact of harpoon fishing on whale sharks led to protective measures.

Challenges and Future Considerations

Balancing economic interests with conservation is a challenge, given the vulnerability of elasmobranchs to fishing pressure. Sustainable management strategies need to consider the relationship between parental stock and recruitment success. Continued research and monitoring are essential to develop effective conservation plans for these ancient and ecologically important species.

Major Order and Family

Hexanchiformes

Overview: Hexanchiformes is an order comprising **seven extant species**, representing the most primitive types of sharks. Fossil evidence suggests similarities to modern **seven gill sharks** dating back to the **Jurassic period**.

Description:

- One spineless dorsal fin positioned over or behind pelvic fins.
- A single anal fin.
- Six or seven gill slits in front of pectoral fins.
- Vertebral column extending into a long dorsal caudal fin lobe, with a small or absent ventral lobe.
- Large mouth, eyes on either side, and small elevated spiracles.
- No nictitating membrane in their eyes.
- Fossil teeth resembling modern hexanchids found in Devonian and Permian deposits, suggesting survival through significant extinction events.

Family: -Hexanchidae (One Genus and One Species Occurs in India)

- Cowsharks, Sixgill, and Sevengill Sharks
- Sharpnosesevengill shark – *Heptranchia sperlo* (Bonnaterre, 1788).
- Bluntnose sixgill shark - *Hexanchus griseus* (Bonnaterre, 1788).
- Bigeye sixgill shark – *Hexanchus nakamurai* (Teng, 1962).

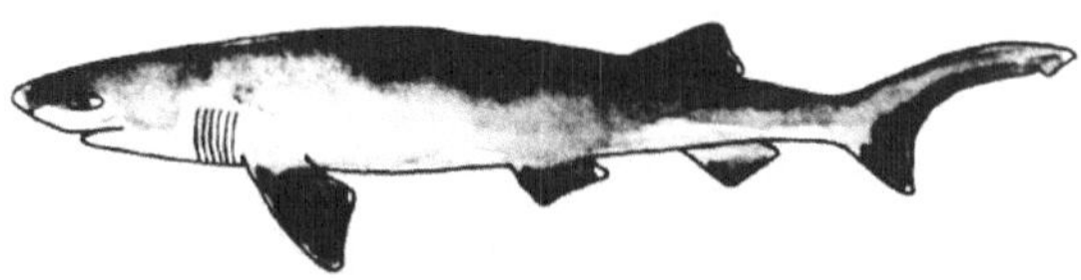

Distinctive Traits

Size and Body: Small to large sharks with varying body shapes, not eel-shaped.

Gill Slits: Head with 6 or 7 pairs of long gill slits, last in front of pectoral-fin origins, first pair not connected across the throat.

Eyes: No nictitating lower eyelids; short, pointed snout; nostrils without barbels or nasoral grooves.

Mouth: Very long mouth extending far behind the eyes; teeth of upper and lower jaws differ, with saw-like lowers having large cusps or cusplets.

Gill Structure: Short dermal gill rakers present on inner gill slits; spiracles present and small.

Fins: Single dorsal fin posterior to pelvic fins; anal fin present; caudal fin less than half total length, strongly asymmetrical with a subterminal notch and a very short lower lobe.

Caudal Peduncle: Not depressed, without keels; no precaudal pits.

Coloration: Grey, blackish, or brown above, lighter below.

Habitat, Biology, and Fisheries of Cowsharks

Habitat: Moderately abundant inshore to deep-water sharks; found in shallow bays to continental slopes and submarine canyons, near or above the bottom.

Diet: Consume a diverse diet, including bony fishes, other sharks, batoid fishes, marine mammals, cephalopods, and crustaceans.

Fisheries: Subject to deep-water line fisheries off Cuba and other regions; incidental catch in trawls. Regularly encountered in targeted shark fisheries and as bycatch in other fishing activities.

Behavior: May snap during capture or when provoked. Larger species encountered by divers are generally docile, with some species becoming the focus of ecotouristic dive trips.

Order - Squaliformes

Squaliformes, comprising approximately **126 species in seven** families.

Characteristics:

- ❑ Two dorsal fins, often with spines.
- ❑ Sharp-headed with no anal fin or nictitating membrane.
- ❑ Five to seven gill slits.
- ❑ Variable in form and size.

Habitat: Typically inhabit saltwater or brackish water.

Distribution: Global presence, from northern to tropical waters, spanning shallow coastal seas to the open ocean.

Family: Echinorhinidae (Bramble Sharks – *Echinorhinus brucus* (Bonnaterre, 1788)

Diagnostic Characters of Large Sharks:

Appearance: Stout cylindrical bodies without abdominal ridges; moderately depressed head.

Gill Slits: Fifth gill slits abruptly expanded; very small spiracles well behind eyes; far-apart nostrils.

Mouth: Short snout; broad, arched mouth; very short labial furrows without encircling; non-papillose lips.

Teeth: Alike in both jaws; strongly compressed, blade-like, with a cusp and up to three side cusplets in adults.

Fins: Two small spineless dorsal fins close together, both smaller than pelvic fins; origin of first dorsal fin behind pelvic-fin origins; anal fin absent; caudal fin without a subterminal notch.

Dermal Denticles: Large and thorn-like.

Color: Blackish to greyish brown above and below.

Habitat, Biology, and Fisheries:

Habitat: Bottom-living on continental shelves and slopes, from close inshore to 900 m.

Diet: Feeds on small bony fishes, other sharks, and crabs.

Fisheries: Not commercially fished in the area but possibly caught as discarded bycatch by offshore demersal fishing fleets; utilized elsewhere.

Distribution:

Range: Wide-ranging in tropical and temperate areas, including the Atlantic, Mediterranean Sea, Indian Ocean, and western Pacific.

Specific Areas: Found in the northern Gulf of Mexico and possibly the Atlantic coast of the USA (known from Massachusetts to Virginia).

Family – Squalidae (Dogfish shark)

- ❑ Shortspine dogfish -*Squalus mitsukurii* Jordan and Snyder, 1903
- ❑ Cuban dogfish - *Squalus cubensis* Howell Rivero, 1936
- ❑ Piked dogfish - *Squalus acanthias* Jordan and Snyder, 1903
- ❑ Rough Skin dogfish – *Cirrhigaleus (Squalus) asper* (Merrett, 1973).

Key Characters

Appearance: Small to moderately large, cylindrical bodies; no ridges between pectoral and pelvic fins; caudal keels and usually precaudal pits on the caudal peduncle.

Gill Slits: Head with 5 gill slits, all anterior to pectoral fins; fifth not abruptly longer; large spiracles just behind eyes.

Snout and Mouth: Short to moderately elongated, broad, flattened snout; no rostral saw; wide-spaced nostrils; short, nearly transverse mouth with smooth lips.

Teeth: Strong-cusped, similar in both jaws, compressed, broad, blade-like, without cusplets; upper teeth nearly as large as lowers.

Fins: Two dorsal fins with a long, strong, ungrooved spine on anterior margins; first dorsal fin larger or subequal to the second; origin far in front of pelvic fins; no anal fin; strongly asymmetrical caudal fin with subterminal notch.

Dermal Denticles: Close-set, not greatly enlarged or plate-like.

Color: Greyish or brownish body and fins without conspicuous marks or luminescent organs.

Habitat, Biology, and Fisheries

Habitat: Spiny dogfishes in deeper water (>50 m) on outer shelves and uppermost slopes in warm-temperate and tropical seas (e.g., Area 31); those in cold-temperate and boreal waters range close inshore and enter the intertidal zone.

Behavior: Often form schools, abundant; feed on fishes and small marine invertebrates; may damage fishing gear.

Significance: Major fishery resource for food and liver oil; intensively fished in many areas as target fisheries and bycatch.

Conservation Concerns: Some species face conservation problems due to overfishing, given their long life, extended maturation periods, and low fecundity.

Order: - Orectolobiformes (Carpet Sharks)

Size: Moderate-sized to gigantic.

Behavior: Sluggish and demersal to active.

Eyes: Without nictitating eyelids.

Facial Features:

- Barbels and nasoral grooves.
- Short mouths ending in front of eyes.
- No enlarged anterior teeth or small intermediate teeth between anteriors and lateral teeth on each side of the upper jaw.

Gill Openings: Five narrow to broad gill openings.

Fins: Two spineless dorsal fins and an anal fin.

Family: Hemiscyllidae (Bamboo Shark)

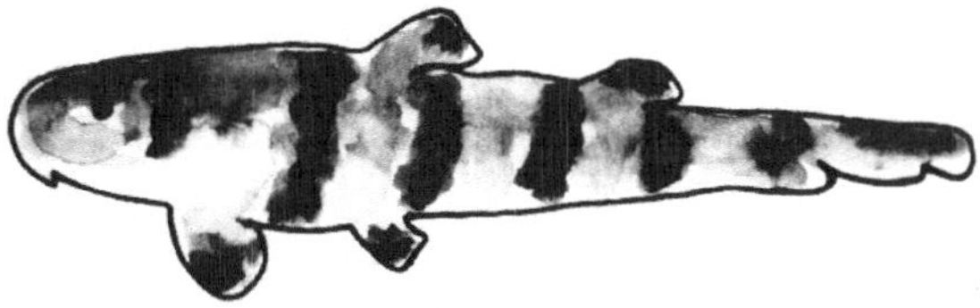

Distinctive Features

Spiracles: Large.

Anal Fin Origin: Well behind the second dorsal-fin origin.

Spacing: Separated from lower caudal-fin origin by a space less than its base length.

Nasal Barbels: Short.

Chin Groove: Absent.

Anal Fin: Low, rounded, and keel-like.

Body Proportions: Distance from vent to lower caudal-fin origin longer than the distance from snout to vent.

Family: Stegostomatidae (Zebra Shark)

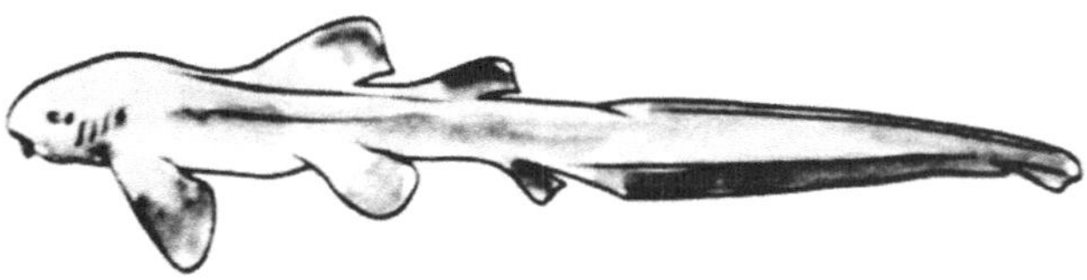

Body: Cylindrical, slightly flattened head, short blunt snout.

Gill Slits: Five short gill slits, last three over pectoral fin bases, fourth and fifth closer together.

Nostrils: Short barbel and groove from each nostril to the mouth.

Mouth: Nearly straight, three lobes on lower lip, furrows at corners.

Teeth: 28–33 tooth rows in upper jaw, 22–32 in lower jaw, with a large central cusp flanked by two smaller ones.

Ridges: Five distinctive ridges in adults, one along dorsal midline, two on sides.

Fins: First dorsal fin halfway along body, twice the size of the second dorsal fin. Large pectoral fins, smaller pelvic and anal fins.

Caudal Fin: Almost as **long as body**, barely developed lower lobe, strong ventral notch near upper lobe tip.

Size: Up to 2.5 m (8.2 ft), with rare reports of 3.5 m (11 ft).

Rare Morph: Sandy zebra shark, overall sandy-brown, lacking distinct dark-spotted pattern.

Family: Ginglymostomatidae

Nurse Sharks (*Ginglymostoma cirratum*)

Color: Yellowish to dark brown.

Fins: Muscular pectoral fins, two spineless dorsal fins (second smaller) aligned with pelvic and anal fins, tail exceeding one quarter of body length.

Mouth: Highly distinctive, subterminal (far ahead of eyes and before snout), reflecting bottom-dwelling nature.

Barbels: Two fleshy barbels on the lower jaw, aiding in prey detection in sediments.

Nostrils: No circumnarial lobe and groove around outer edges of nostrils

Eyes: Mouth far ahead of eyes, emphasizing their benthic (bottom-dwelling) behavior.

Spiracles: Small, circular openings behind each eye as part of the respiratory system.

Teeth: Serrated, fan-shaped, continuously replaced throughout the shark's life.

Rhincodontidae (Whale Shark)

Mouth: Huge and nearly terminal (at the end of the snout).

Gill Slits: External gill slits are very large, and there are internal gill slits inside the mouth cavity with filter screens.

Caudal Peduncle: Strong lateral keels.

Caudal Fin: Strong ventral lobe, lacking a strong terminal lobe and sub-terminal notch.

Lamniformes

Body Shape: Typically have a fusiform (spindle-shaped) body, well-adapted for swift swimming.

Gill Slits: Usually possess five gill slits, with a few exceptions.

Teeth: Sharp, often serrated teeth, with well-developed upper and lower jaws.

Dorsal Fins: Two dorsal fins, each with a spine, commonly present.

Tail: Heterocercal tail, with the upper lobe larger than the lower lobe.

Anal Fin: Typically, absent.

Nictitating Membrane: Well-developed nictitating membrane over the eyes.

Habitat: Inhabit a variety of marine environments, from coastal to open ocean.

Alopidae – Theresher Sharks

Body Structure: Large size; cylindrical trunk and precaudal tail, not depressed, and lacking lateral ridges.

Head: Not laterally expanded or depressed; eyes on the sides without nictitating lower eyelids.

Snout: Moderately long, bluntly conical, not flattened, without lateral teeth or barbels; nostrils well-separated from the mouth.

Mouth: Small but arched and elongated, extending well behind the eyes; labial furrows present on the lower jaw or absent.

Teeth: Small, blade-like, compressed, with erect to oblique cusps; anterior teeth in upper jaw slightly larger than lateral teeth.

Gill Slits: Five small to medium-sized gill slits, last 2 behind pectoral-fin origins; no gill rakers on internal gill slits; minute spiracles present.

Fins: Two dorsal fins without spines; first dorsal fin moderately large, high, angular, and much shorter than the caudal fin; second dorsal fin low and minute; anal fin present, very small.

Caudal Fin: Strongly asymmetrical, upper lobe enlarged, about half the total length with a subterminal notch, undulated dorsal margin; lower lobe short but strong.

Caudal Peduncle: Not depressed, without keels; precaudal pits present

Family: Lamnidae : (Mackerel Shark)

Mackerel Sharks (Makos, White Sharks, Porbeagles)

Size and Shape: Large-sized sharks with a fusiform body.

Gill Slits: Head with 5 gill slits, all in front of pectoral-fin origins; no nictitating eyelids.

Teeth: Long and few in number, awl- or blade-like, with a single cusp; gill arches lack rakers.

Dorsal Fins: Two dorsal fins; the first dorsal fin much shorter at the base than the caudal fin, positioned far in advance of pelvic fins.

Second Dorsal and Anal Fins: Much smaller than the first dorsal fin, with narrow, pivoting bases.

Caudal Fin: Lunate, less than 1/3 of total length; caudal peduncle strongly depressed dorsoventrally, expanded laterally, with prominent keels on each side extending well out on the caudal fin.

Intestinal Valve: Ring type.

Color: Back in various shades (pale grey, greyish blue, purplish blue, brownish, blackish grey, or black); underside white to lighter grey.

Order:- Carcharhiniformes (Ground Sharks)

Trunk: Cylindrical to slightly compressed or depressed but not raylike.

Head: Conical to depressed, usually not anteriorly expanded. In Sphyrnidae, prebranchial head is present.

Gill Slits: 5 pairs on the sides of the head, with the last 1 to 3 over pectoral bases; some Scyliorhinidae have dorsolateral gill slits.

Spiracles: Present in many species, varying in size, located close behind eyes, or absent.

Nostrils: Usually without barbels and nasoral grooves; circumnarial grooves absent. Barbels, if present, develop from anterior nasal flaps.

Eyes: Lateral or dorsolateral, with true nictitating lower eyelids.

Snout: Short to moderately long, not greatly elongated, not forming a rostral saw with lateral teeth and barbels.

Mouth: Moderately large to very large, arched, extending behind anterior ends of eyes; labial furrows varying.

Teeth: Usually without enlarged molariform posterior teeth; anterior teeth not separated by small intermediate teeth or a gap from lateral teeth.

Dorsal Fins: Two dorsal fins (possibly only one in Pentanchus, family Scyliorhinidae), without spines.

Pectoral Fins: Moderate-sized to large, not raylike, without triangular anterior lobes.

Pelvic Fins: Small to moderately large, vent continuous with inner margins.

Caudal Fin: With a long dorsal lobe, ventral lobe varying from long to absent; vertebral axis raised into the dorsal caudal lobe.

Intestinal Valve: Spiral or scroll type.

Family: Scyliorhinidae:(Cat Sharks)

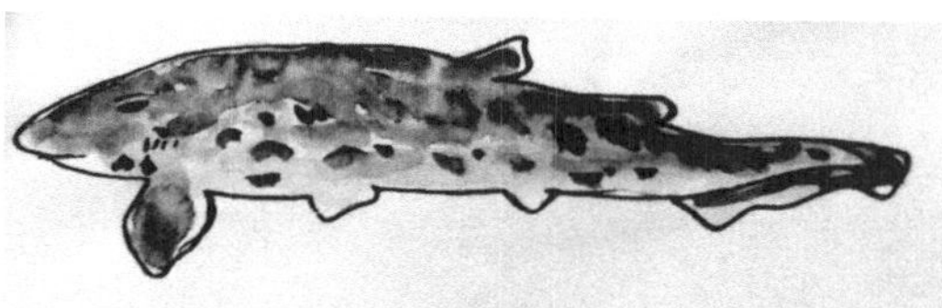

Head: Without laterally expanded blades; elongated and fusiform, oval, or slitlike eyes, length over 1.5 times the height; rudimentary nictitating eyelids; moderately large spiracles.

Nasal Flaps: Anterior nasal flaps variably shaped, not barbel-like (except for Poroderma genus).

Nostrils: Internarial width about 0.6 to 1.3 times nostril width.

Labial Furrows: Absent or very short to very long.

Teeth: Small, with acute narrow cusps, often with lateral cusplets and basal ledges; not bladelike; similar in both jaws; posterior teeth comblike or not; tooth rows range from 40 to 101/34 to 111.

Precaudal Pits: Absent.

Dorsal Fins: First dorsal fin small, not keel-like, much shorter than the caudal fin; positioned over or behind pelvic bases, its origin slightly ahead of pelvic origins in some cases.

Pectoral Fins: Radials confined to bases of fins.

Caudal Fin: Ventral caudal lobe absent or very weak; no undulations or ripples in dorsal caudal margin.

Neurocranium: May or may not have a supraorbital crest.

Vertebral Centra: May or may not have strong, wedge-shaped intermedial calcifications.

Reproduction: Usually oviparous, but some species are ovoviviparous.

Family: Proscyllidae (Finback catsharks)

Precaudal Pits: Absent.

Dorsal Caudal Margin: Not undulated.

Teeth: Posterior teeth on dental bands are comblike.

Labial Furrows: Very short or absent; if present, confined to mouth corners.

Family: Leptocharidae (Barbeled Hound Sharks)

Anterior Nasal Flaps: Formed as slender barbels.

Labial Furrows: Upper labial furrows extremely long, nearly equal to internarial width and over half mouth width.

Intestinal Valve: With 14 to 16 turns.

Neurocranium: No supraorbital crests.

Family: Triakidae (Hound Sharks)

Anterior Nasal Flaps: Usually not barbel-like (except for Furgaleus).

Labial Furrows: Upper labial furrows shorter, considerably less than internarial width and less than half of mouth width.

Intestinal Valve: With 4 to 10 turns.

Neurocranium: Supraorbital crests present.

Family: Hemegaleidae (Weasel Shark)

- **Anterior Nasal Flaps**: Usually not barbel-like.
- **Labial Furrows:** Upper labial furrows shorter, considerably less than internarial width, and less than half of mouth width.
- **Intestinal Valve:**Typically with 4 to 10 turns. Intestinal valve of spiral type
- **Neurocranium:** Supraorbital crests present.

Family: Carcharhinidae (Requiem sharks)

Intestinal valve: Scroll type

- **Distribution:** Global. Gill openings 5, the fifth behind origin of pectoral fin.
- Small to large sharks with round eyes, internal nictitating eyelids, no nasoral grooves or barbels, usually no spiracles.
- Teeth usually bladelike with one cusp. Development usually viviparous with young born fully developed. Includes several dangerous species, but most prefer to avoid divers.

Family: Spirinidae – (Hammerheaded Sharks)

Head with lateral, bladelike expansions

- Marine, (coastal;) occasionally in brackish water.
- (**Distribution:** Global chiefly warm waters).
- Head laterally expanded, with eyes and nasal openings much wider set than in other sharks. The blades presumably server to increase the sharks sensory capabilities.
- No spiracle.

Chapter-9

"Batoid Fishes Diverse Morphology and Ecological Significance in Marine Ecosystems" (Saw Fish, Skates and Rays)

- **Body Shape**: Disc-shaped, rounded, or sub-angular; flattened body with fused pectoral fins along the sides of the head.
- **Eyes and Mouth:** Eyes are positioned superiorly (on the top side of the body), while the mouth is inferior (on the underside); mouth is often protractile.
- **Gill Slits:** Typically, there are five gill slits located on the underside of the body.
- **Spiracles**: Presence of spiracles, which are modified gill slits located behind the eyes on the top side of the body; these spiracles play a crucial role in respiration.
- **Dorsal Fins and Tail:** If dorsal fins are present, they are usually located on the tail; no anal fin; tail is typically reduced, sometimes appearing as a filament.
- **Embryonic Development:** In embryonic stages, some show shark-like affinities before developing the characteristic fused pectorals along the head.

- These batoid fishes, including rays, skates, sawfishes, and guitarfishes, are primarily bottom-dwelling species found in various habitats. Their spiracles are crucial for respiration, drawing water in through the gills and expelling it through the spiracles. Many of these species are of commercial importance as food sources.
- Their flattened body shapes and adaptations make them well-suited for life on the seafloor, where they feed on a variety of prey. Sawfishes, for instance, have elongated snouts lined with teeth that resemble a saw, which they use to detect and catch prey.

Habitats and Distribution

Ecological preferences: Where these species are typically found in coastal regions, deep sea, specific oceanic zones, etc.

Global distribution: Different regions worldwide where various batoid species are prevalent.

Order: - Rajiformes

Family: Rajidae (Skates)

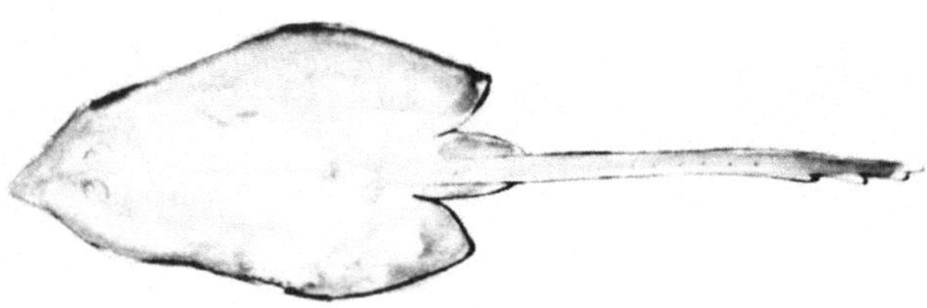

Body Shape

- Rounded or angular body.
- Laterally widened body, which can be interpreted as greatly flattened from side to side.

Body Structure

- Body and head greatly depressed, forming a rhomboidal disc.
- Pectoral fins fused with the body, contributing to the disc shape.

Tail Structure

- Tail terminates in a blunt tip without a caudal fin.

Fins

- Presence of two dorsal fins located posteriorly (toward the rear) on the body.
- Dorsal fins are distinct and relatively small.
- They are positioned near the end of the caudal (tail) region.

Examples

- ❑ ***Raja clavata*** - **Common skate or thornback skate**: Found in the northeastern Atlantic Ocean and the Mediterranean Sea.
- ❑ ***Raja radiata*** - **Clearnose skate**: Found in the western Atlantic Ocean, ranging from North Carolina to Brazil.
- ❑ ***Raja eglanteria*** - **Roughtail skate**: Found in the western Atlantic Ocean, particularly in the Gulf of Mexico and along the southeastern coast of the United States.
- ❑ ***Raja fullonica*** - **Fullback skate**: Native to the Mediterranean Sea and the eastern Atlantic Ocean.

Rhinobatidae (Guitar fishes)

- ❑ The Rhinobatidae family, commonly known as guitarfishes, comprises a group of cartilaginous fishes that possess a mix of characteristics akin to both rays and sharks.

Body Shape

- ❑ Rather elongated body compared to other rays and skates.
- ❑ The snout is typically pointed and elongated.

Snout Structure

- ❑ Broad, soft, and rounded snout, which is a distinctive feature of guitarfishes.
- ❑ The pointed snout is notably different from the flattened disc-shaped heads of typical rays.

Dorsal Fin

- ❑ The first dorsal fin is triangular in shape.
- ❑ Positioned anteriorly, usually before the base of the pelvic fins.

Eyes and Spiracles

- ❑ Eyes are slightly smaller in proportion to the body and are entirely separated from spiracles.
- ❑ This separation between the eyes and spiracles distinguishes them from some other batoids where the eyes and spiracles might be more closely positioned.
- ❑ Some common species within the Rhinobatidae family include the shovelnose guitarfish (*Rhinobatos productus*) found along the Pacific coast

of North America and the whitespotted guitarfish (*Glaucostegus typus*) found in the Indo-Pacific region.

Order:- Rhinopristiformes

Family: Pristidae (Saw Fishes)

Key Characteristics

Body Shape

- Elongate body, not resembling the typical flattened disc shape of most rays.
- The body often appears slender and elongated compared to other batoids.

Snout Structure

- The snout is elongated and produced.
- It is notable for its saw-like structure lined with teeth.
- The teeth along the rostrum are bony, prominent, and arranged in a saw-like pattern.

Rostral Teeth

- The teeth at the posterior end of the rostrum extend well beyond the anterior two pairs of rostral teeth.
- These teeth are a defining feature of sawfish and are used for various functions, including hunting and defense.
- Several species of sawfish exist, including the smalltooth sawfish (*Pristis pectinata*) and the largetooth sawfish (*Pristis pristis*), among others.

Torpediniformes

Family: Torpedinidae (Electric rays)

Body Shape

- Typically, rounded or angular body.
- The body is laterally widened, often forming a disc-like shape.

Electric Organs

- Notably, electric rays possess two large electric organs located in the front part of the disc, one on each side of the head.
- These organs generate electric charges that the ray uses for hunting, defense, and communication.

Eye Size and Coloration

- The eyes of electric rays are relatively small compared to the size of their disc-shaped body.
- The dorsal surface often displays spots against a brown background, while the ventral (lower) side tends to be pale in coloration.
- *Narcine timlei*, commonly known as the Timly Electric Ray, is a species within the Torpedinidae family.

Narcinidae (Numb fishes)

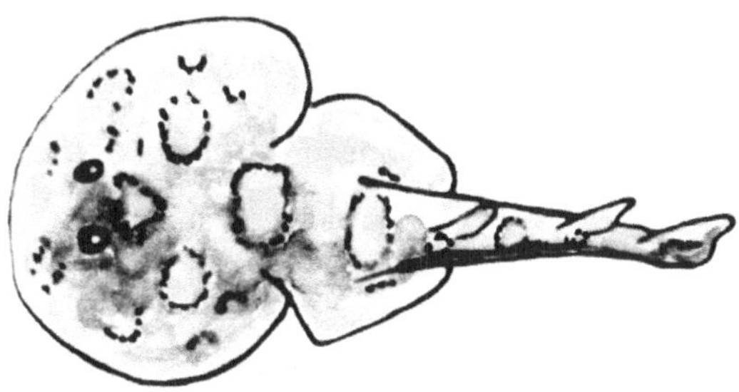

Size and Body Shape

- Small to medium-sized batoids, typically between 15 to 66 cm in total length, often less than 50 cm.
- Large oval, rounded, or shovel-shaped pectoral discs.
- Stout shark-like tails; tail either equal to or longer than the disc length.

Skin and Body Structure

- Naked body without dermal denticles or thorns, both above and below.
- Depressed and flattened trunk, distinct from a shark-like body structure.
- Precaudal tail moderately depressed, sometimes with lateral ridges on sides, and abruptly narrower than the trunk.
- No barbed sting or stinging spine on the dorsal surface of the tail.

Head and Snout

- Broad and depressed head with a moderately elongated snout, which is broadly rounded or obtusely rounded-angular.
- Large rostral cartilage supporting the snout, without forming a saw-like structure or lateral saw teeth.

Gills and Mouth

- Five small gill openings on the underside of the front half of pectoral-fin bases.
- Dorsolateral eyes just anterior and partly medial to spiracles.
- Mouth transverse and straight with strong labial folds and a prominent groove around its periphery.
- Nostrils located just anterior to the mouth, connected by broad nasoral grooves.

Teeth

- Small oral teeth with a rounded or oval base and short to moderately long cusps, not laterally expanded.
- Tooth bands remain exposed when the mouth is closed, rounded to angular in outline, and between 12 and 33 exposed rows in either jaw (in adults).

Fins

- Pectoral fins originating at the anterior edges of the snout and attached to most of the head.
- Pelvic fins are low, subangular or rounded, not divided into anterior and posterior lobes.
- Two moderately large and widely separated dorsal fins present, similar in angular or subangular shape.

Coloration

- Varied dorsal surface colors, ranging from whitish, yellowish, brownish, to black, often with spots or blotches. Usually, the ventral surface is white but can be black in deep-water species.

Examples:-

- ***Narcine sbancroftii*** - Also known as the lesser electric ray or the Panama electric ray. It's found in the Eastern Pacific Ocean from Mexico to Peru. This species showcases the typical features of the Narcine genus, including the electric organs and rounded snout.
- ***Narcine brasiliensis*** - Known as the Brazilian electric ray. It inhabits the western Atlantic Ocean, ranging from the Caribbean Sea to southern Brazil.

- ❑ ***Narcine entemedor*** - The numbfish or Atlantic electric ray, found in the western Atlantic Ocean, particularly in the Gulf of Mexico and the Caribbean Sea.

Order: Myliobatiformes

Family: Gymnuridae (Butterfly rays)

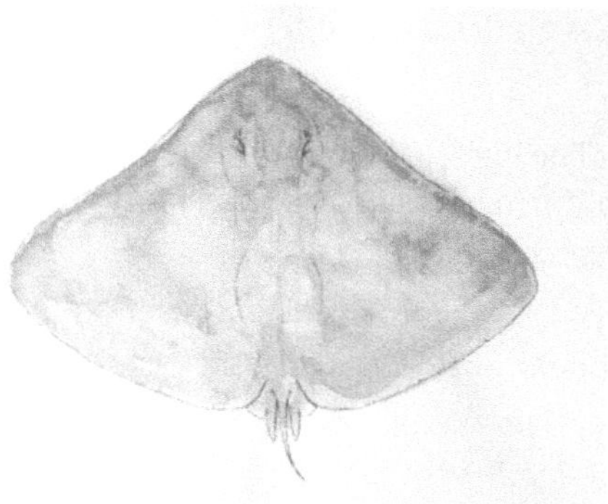

- ❑ Medium to large-sized stingrays, with a disc width exceeding 2 meters.
- ❑ The body is strongly depressed, forming a rhomboid disc shape with the head, trunk, and expanded pectoral fins.
- ❑ The disc is notably broader than it is long, at least 1.5 times wider than long.
- ❑ The family Dasyatidae includes various stingrays, such as the southern stingray (*Dasyatis americana*), the spotted eagle ray (*Aetobatus narinari*), and other species that possess these typical characteristics. s

Family: Dasyatidae

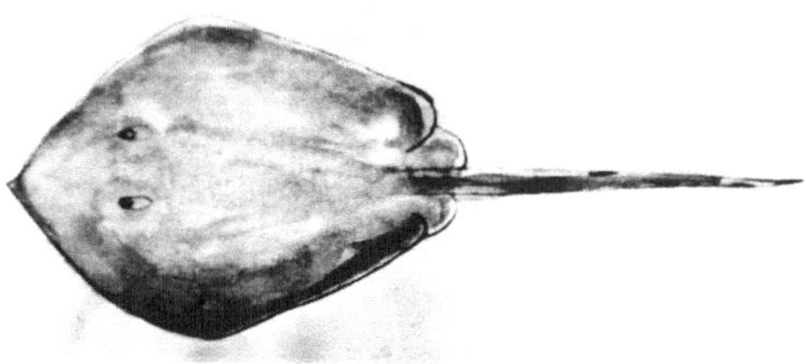

Body Shape

- ❑ Rounded or angular body, laterally widened.
- ❑ The disc is at most 1.3 times as broad as long.
- ❑ Tail significantly longer than the disc width.

Mouth Structure:

- ❑ Mouth floor with several fleshy papillae (protruding, nipple-like structures).

Fins and Tail

- Caudal fin thin, and the dorsal fin may be feeble or absent.
- One or two serrated spines present in the tail.

Snout and Head

- Normal snout, not notably elongated or modified.
- Head is not elevated, maintaining a relatively flat position.

Examples

Raja binoculata - The big skate: Found along the Pacific coast of North America, from Alaska to Baja California. It has a rounded body shape and a relatively slender tail compared to its disc width.

- ***Raja radiata*** - The thornback ray: Common in the Eastern Atlantic and the Mediterranean Sea, characterized by a rounded disc and a slender tail.
- ***Raja clavata*** - The thornback skate: Inhabits the Northeastern Atlantic Ocean and the Mediterranean Sea, displaying a rounded or angular body and a tail longer than its disc width.

Family: Myliobatidae (Eagle rays and Cownose rays)

Body Shape

- Rounded or angular body, laterally widened.
- Typically, a more pronounced disc shape with a wider span.

Fins and Tail

- Caudal fin thin, and the dorsal fin is often feeble or entirely absent.
- Their tails are whip-like and notably longer than the length of their bodies.

Head and Body Structure

- The head is elevated above the pectoral fins, giving a somewhat raised appearance to the front of the ray's body.
- This elevation creates a distinctive structure where the head seems lifted above the main body, distinguishing them from some other ray species.

Examples of ray species within the Myliobatidae family include the spotted eagle ray (*Aetobatus narinari*), the cownose ray (*Rhinoptera bonasus*), and the bullnose ray (*Myliobatis freminvillii*).

Family :Mobulidae (Devil rays and Manta rays)

Body Shape

- Rounded or angular body, laterally widened.
- The body form is typically broader.

Pectoral Fins

- The pectoral fins are highly enlarged and wing-like, contributing to their distinct appearance.

Fins and Tail

- Similar to other rays, they have a thin caudal fin, and the dorsal fin is typically feeble or absent.

Snout and Cephalic Organs

- The snout is produced as a fleshy flap on each side, which is a distinguishing feature.
- These rays have distinct modified cephalic (head) organs that are often present in pairs and contribute to their unique sensory capabilities.

Examples

- Within the Mobulidae family, examples include the giant manta ray (*Manta birostris*) and the lesser devil ray (*Mobula hypostoma*), both of which showcase these distinctive characteristics.

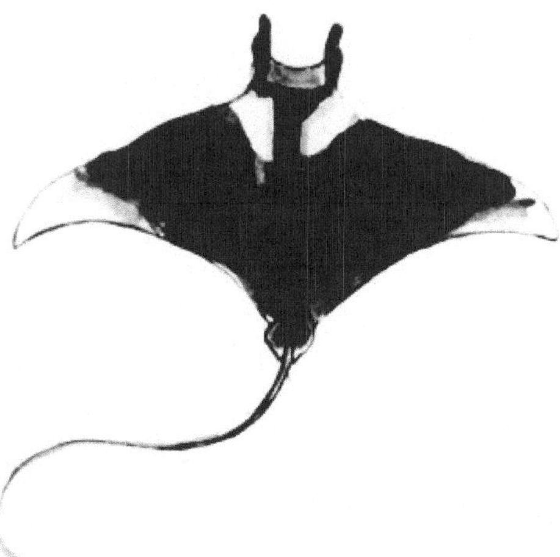

Chapter-10

Teleostomes

Teleostomi

- The Rise of Jawed Fishes: Devonian Era and the "Age of Fishes"
- The Devonian period stands as the "age of fishes," witnessing the flourishing of diverse freshwater and marine fish groups. This era marked the emergence of jawed fishes within the fossil record. Among these, the acanthodians played a pivotal role, being linked to both cartilaginous and bony fishes. Their evolutionary significance paved the way for the classification of Teleostomi.
- Acanthodians and the Foundation of Teleostomi.
- Distinctive Mouth Mechanism: Notably, the evolution of jawed fishes introduced a unique mouth-opening mechanism. This involved the downward movement of the mandible through the hyoid apparatus, characterizing the Teleostomi group.

Osteichthyes: Dominant Bony Fishes of the Devonian Era

During the Devonian period, Osteichthyes emerged as dominant due to their efficient movement and feeding mechanisms. This group further divides into:

Sarcopterygii and Actinopterygii

Sarcopterygii: Inhabited freshwater bodies with lobed fins facilitating resting and locomotion.

Actinopterygii: Known for streamlined fins and active swimming, marked by the presence of bones.

Within Sarcopterygii, Romer classified: Dipnoi and Crossopterygii

- **Dipnoi (Lungfishes):** Flourished during the Permian and Triassic eras but became extinct afterward, surviving today in three genera across separate geographic regions.
- **Crossopterygii:** Adapted to marshy environments, contributing to the evolution of terrestrial tetrapods.

Crossopterygii: Transitional Fishes

- **Rhipidistia:** Extinct group
- **Actinistia (Coelacanthiformes):** ***Latimeria chalumnae*** **:Rediscovery of a "Living Fossil"**
- Initially presumed extinct for over 50 million years, Latimeria chalumnae was discovered in 1938 off the South African coast.
- Typically steel blue and sizable, weighing 57.5 kg and measuring 1.5 m.
- Multiple specimens found between Madagascar and Mozambique at depths of 250-300 m.

Coelacanths: Living Fossils

- The discovery of living coelacanths in 1938 was monumental, challenging the belief that they'd been extinct for over 50 million years. These creatures, with minimal evolutionary change, provide a window into ancient vertebrate anatomy and hint at the existence of other primitive sea life.

Phylogeny and Anatomical Resemblances

- **Chondrichthyes Similarities:** High urea in blood, pituitary gland resemblance.
- **Bony Fish Resemblance**: Bony scales, swim bladder, and skeletal similarities.
- **Unique Attributes:** Possess distinctive features, making their evolutionary connection uncertain.

Dipnoi: Lung Fishes

Primitive Traits and Features:

- **Ancient Characteristics:** Persistent notochord, cartilaginous skeleton, and cloaca presence.
- **Resemblance to Primitive Teleostomi**: Lobate fins, cosmoid scales, and nostril location.
- **Similarities with Crossopterygii:** Lobate fins, diphycercal tail, and scale characteristics.
- **Amphibian Resemblance:** Dual respiratory organs, aestivation, and certain physiological parallels.

Classification and Distribution

Subdivisions: Monopneumona (one lung) and Dipneumona (two lungs) based on lung presence and fin structure.

Geographical Discontinuity: Limited distribution across Africa, South America, and Australia.

Life Cycle and Habitat

- **Behavioral Traits:** Diverse behaviors, from respiration strategies to nesting habits.
- **Neoceratodus:**Orgin-Australia, 4 pairs of gill, 1 lung present. Bottom-dweller, shifts between aquatic and aerial respiration, and complex spawning rituals.

- **Protopterus:**Orgin-South Africa, 4 pairs of gill, 2 lungs present. Marsh inhabitant, heavily relies on lungs for respiration, and shows aestivation.

- **Lepidosiren:**Orgin-South America, 5 pairs of gill slites, 2 lungs present. Feeds on mollusks, uses both gills and lungs for respiration, and exhibits nesting behaviors.

Evolutionary Significance

Transitional Group: Bridges fish and amphibians, indicating a potential evolutionary relationship.

Unclear Connections: Some similarities might be adaptive rather than evolutionary.

Implication as Tetrapod Ancestors: Offers insights into vertebrate evolution without confirming a direct lineage. Dipnoi stand as an enigmatic group, showcasing primitive and specialized traits. Their existence provides a glimpse into the evolutionary shift between fish and amphibians, hinting at a shared ancestral origin. Dipnoi's significance lies in their ancient nature, serving as a window into vertebrate evolutionary history.

Chapter-11

Actinopterygii: Higher Bony Fishes

Introduction

- Actinopterygii encompasses higher bony fishes.
- Sub-class Actinopterygii constitutes all living ray-finned fishes.

Key characteristic features:

- Fins supported by fin rays, with reduced bones.
- Thin, horny scales - cycloid or ctenoid, excluding cosmoid; some primitive forms have ganoid scales, while others lack scales.
- Generally homocercal tail; some primitive species have a heterocercal caudal fin.
- Single external branchial aperture.
- Presence of operculum.
- Absence of spiracle.

Classification Challenges

Historical classification:

- **Initially Grouped into three Superorders:** Chondrostei, Holostei, Teleostei.
- Regan (1929) proposed Palaeopterygii and Neopterygii.

- Berg (1940) argued against the artificiality of the Chondrostei, Holostei, Teleostei division, proposing division into multiple orders.

Orders of Actinopterygii

Palaeonisciformes

- Ancient fishes like "Palaeoniscids."
- Existed from the Devonian to the Triassic.
- Ancestral to all actinopterygians - *e.g.*, Cheirolepis.

Polypteriformes

- Ten species of African bichirs and reed fishes.
- **Characteristics:** well-ossified skeleton, thick multilayered ganoid scales, symmetrical caudal fin, presence of spiracle.

Example: *Polypterus bichir* - a predator feeding on crustaceans, small fishes, and amphibians.

Acipenseriformes

- Includes sturgeons (Acipenseridae) and paddle fishes (Polyodontidae).
- **Sturgeons:** elongated body, five rows of bony scutes, heterocercal tail, protrusible jaws with barbels.
- **Paddle fish *(Polyodon spatula)*:** greatly elongated rostrum, absence of bony scutes, feeds by filtering crustaceans and small fishes.

Amiiformes

- Holostean fishes like gar-pike (Lepisosteus) and Bow-fin (Amia).
- **Characteristics:** ganoid scales, symmetrical tail, absence of spiracles.

Example: Lepisosteus (gar-pike) - predatory, with primitive and specialized traits.

The Teleosts

- Most living freshwater and marine fishes.
- Evolved in the sea, later adapted to fresh water.
- Extensive diversification in the Cretaceous with various adaptations.

Infra Divisions of Teleosts

Osteoglossomorpha (Bony Tongues)

- Represented by Osteoglossum and Mormyrus.
- Adaptations for living in specific environments - *e.g.*, *Proboscis*-like extension for probing mud in Mormyrus.

Elopomorpha

- Includes tarpons, ladyfish, eels - characterized by elongated eel-like bodies, some without scales.

Clupeomorpha

Consists of herrings, sardines, anchovies, adapted for feeding on plankton.

Euteleostei

- The majority of living teleosts, divided into further superorders: Ostariophysi, Protacanthopterygii, Paracanthopterygii, Acanthopterygii.

Ostariophysi

- Characterized by Weberian ossicles connecting the inner ear to the swim bladder, enhancing sound sensitivity.
- Predominantly freshwater fishes, forming the base of euteleostean radiation.

Examples

- **Family Cyprinidae:** Labeo spp., Cirrhina spp., Catla, Cyprinus, Puntius, etc.
- **Family Cobitidae:** Loaches like Noemacheilus, *Botiaalmorhae*, *Lepidocephalichthys guntea*.
- **Family Siluridae:** Catfishes such as *Wallago attu,Ompok bimaculatus*, *Clarias batrachus*, etc.

Protocanthopterygii

- Esocid and Samnid game fishes are included in this taxon.
- Includes ancestral euteleostean characters seen in salmon species, trout, and deep-sea lantern fishes having photophores on their body.

Paracanthopterygii

- Mobile jaws and protective spines in the median fins
- Ex. Cods and angler fishes

Acanthopterygii

- Truly rayed fishes inhabiting shallow marine, freshwater bodies includes about 9000 species of Perciformes, flying fishes, half-cyprinodonts, and swordtails.
- **Examples:** Exocoetus, Hemiramphus, Xenentodon, Aplocheilus, Gambusia, etc.

Chapter-12

Order: Clupeiformes

Order - Clupeiformes (Majority of species found in Marine water but few species Brackish water and Freshwater)

- ❑ Having a single dorsal fin in the middle
- ❑ Only soft rays, fin spines are present.
- ❑ Pelvic fins in abdominal position
- ❑ Lateral line absent
- ❑ Usually a row of scutes along ventral profile.

Kingdom: Animalia

Phylum: Chordata

Class: Actinopterygii (ray-finned fishes)

Order: Clupeiformes

Family: Clupeidae, Engraulidae and Chirocentridae.

No	Family Name	Key characters	Other features
1.	**Clupeidae** (Herrings, Sardines, Shads and Menhadans - 7 genus and 15 species). **(a) Subfamily: Dussumieriinae (Round Herring)** 1. **Rainbow sardine** -*Dussumeeria acuta* (Valenciennes) **(b) Subfamily: Spratelloidinae** 1. **Silver stripe round herring** - *Sparatelloides gracilis* (Schlegel) 2. **Blue sprat (Delicate round herring)** - *S. delicatulus* (Bennett) **(C) Subfamily:** Clupeinae 1. **White sardine**- *Escualosa thoracata (Valenciennes)* 2. **Spot back herring** -*Herklotsichthys punctatus (Ruppell)* 3. **White sardinella**- *Sardinella albella (Valenciennes)* 4. **Deep body sardine**- *S. brachysoma* Bleeker 5. **Smooth belly White sardine** *S. Clupeiodes* (Bleeker) 6. **Sind sardine** *S. dayi Regan*	**Body Shape and Characteristics:** • Body shape varies from fusiform (spindle-shaped) to either round or strongly compressed. • Lack of scales on the head. • Jaw teeth, if present, are typically small or minute. • Belly scutes present. **Fin Structure:** • Single dorsal fin, usually small and positioned near the midpoint of the body. • Pelvic fins are generally located more or less below the base of the dorsal fin. • In some species, dorsal and pelvic fins may be absent. • Fins consist of soft rays only. **Lateral Line and Scales:** • Lateral line may extend a few scales behind the head in some species, but it might be absent in others. • Scales are typically cycloid, meaning they are smooth to the touch. **Branchiostegal Rays:** • Usually, there are 5-10 branchiostegal rays. 	• **Distribution:** global (mostly tropical) from 70° N to about 60° S. • Chiefly marine coastal and schooling fishes; some freshwater and anadromous. • Most feed on small planktonic animals. • **Size range (adults):** from 2 to 75 cm. • One of the most important family of commercial fishes, processed for food, oil, or fish meal.

	7. **Fringe scale sardine** - *S. fimbriatta* (Valenciennes) 8. **Gold stripe sardine** - *S. gibbossa* Bleeker 9. **Blue sardine** *S. leiogaster* (Valenciennes) 10. **Indian oil sardine** - *S. longiceps* (Valenciennes) 11. **Black tip sardine** - *S. melanura* (Cuvier) 12. **Spotted(Pilchard) sardine** *S. sirm* (Walbaum). **3. Subfamily: Pristigasterinae** 1. **Chinese herring (Slender shad)** - *Ilisha elongatta* (Bennet) 2. **Coramandel ilisha** - *I. filigera* (Valenciennes) 3. **Kempen's ilisha** - *I. kempeni* (Weber and de Beaufort) 4. **Bigeye ilisha**- *I. megaloptera* (Swainson) 5. **Indian ilisha** - *I. melastoma* (Schneider) 6. **Lobejaw ilisha** - *I. sirishai Seshagiri*- Rao	**4. Subfamily: Alosinae (Shads)** 1. **Hisha shad (Herring)** - *Hilsa ilisha* (Hamilton) 2. **Five spot herring (Razor belly)** - *H. kelee* (Cuvier) 3. **Chinese herring** *H. toli* (Valenciennes) 4. **Indian river shad** - *Gadusa chapra* (Hamilton) 5. *G. suhia* (Chaushuri **5. Subfamily: Pellonulinae** 1. **Day's round herring** *Dayella malabarica* (Day) 2. **Ganges river sprat** -*Corica soborna* Hamilton 3. **Malabar sprat** - *Ehirava fluviatilis* Deraniyagala **6. Subfamily: Dorostomatinae** 1. **Ganges river gizzard shad** *Gonialosa manmina* (Hamilton) 2. **Thai gizzard shad**- *Anodontosoma chacunda* (Hamilton) 3. **Galatheae gizzard shad** *Nematalosa galatheae* (Nrlson and Rothman) 4. **Hair back (Gizzar shad, Long-ray bony bream)** *N. nasus* (Bloch)	

	7. **Tapra fish** - *Ophisthopterus tardoore* (Cuvier) 8. **Indian pellona** - *Pellona ditchela* Valenciennes 9. **Raconda** - *Raconda russeliana* (Gray)		
2.	**Engraulidae** **1. Subfamily: Collinae** 1. **Gold spot grenadier anchovy** - *Colia dussumieri* Valenciennes. 2. *C. korua Dutt & Seshagiri-Rao* 3. **Osbeck's grenadier anchovy** - *C. mystus* (Linnaeus) 4. **Neglected grenadier anchovy** - *C. neglecta* Whitehead 5. **Ramacarat grenadier anchovy** - *C. ramacarati* (Hamilton) 6. **Reynald grenadier anchovy** - *C. reynaldivalenciennes* **2. Subfamily: Engraullinae** 1. **Common hairfin anchovy** - *Setipinna Godavari* (Babu-Rao) 2. **Gangetic hairfin anchovy** - *S. phasa* (Hamilton) 3. **Scaly hairfin anchovy** - *S. taty* (Valenciennes) 4. **Andhra (Spiny) anchovy** - *Stolephorus andhraensis* (Babu-rao)	**Morphological Traits:** • **Size:** Typically, small to moderate-sized (10 to 20 cm standard length, occasionally larger). • **Snout:** Prominent pig-like projection beyond the lower jaw tip. • **Jaws:** Slender, long lower jaw extending past the eye. • **Teeth:** Usually small; in Lycothrissa, they're canine-like. • **Eyes:** Large, covered by an adipose eyelid. • **Gill Rakers:** Short and not numerous in Indo-Pacific genera. **Skeletal Features:** • **Pelvic Scute:** Always has ascending lateral arms. • **Scutes:** Indo-Pacific species often have pre- and postpelvic scutes and a spine-like scute anterior to the dorsal-fin origin. • **Fins:** Single, short dorsal fin near body midpoint. No adipose fin. Pectoral fins set low. Pelvic fins with specific soft ray counts. • **Anal Fin:** Moderate length, varying in soft fin ray count.	• **Marine Coastal:** Predominantly found in marine coastal regions. • **Wide Range:** Occur in seas from about 60°N to 50°S latitudes. • **Adaptability:** Some enter brackish or freshwater for feeding/spawning, while others permanently reside there. • **Feeding:** Primarily feed on small planktonic animals, especially crustaceans. • **Reproduction:** Scatter large numbers of eggs; larvae hatch into planktonic forms. • **Fisheries Importance:** • **Commercial Significance:** Crucial in fisheries, contributing up to 25% of total clupeoid catch. • **Regional Contribution:** Certain species contribute significantly, up to 6% in specific catches. • **Staztistics:** Detailed data often lacking, but some species, like Thryssa, may warrant specific reporting.

<table>
<tr>
<td></td>
<td>5. Common hairfin anchovy - S. bataviensis Hardenberg
6. Buccaneer anchovy - S. buccaneeri Strasburg
7. Spined anchovy - S. tri (Bleeker)
8. Baelama anchovy -Thrissina baelama (Forsskal)
9. Dussumier's anchovy -T. dussumieri (Valenciennes)
10. Hamilton's anchovy T. hamiltoni (Gray)
11. Kammal anchovy - T. kammalensis (Bleeker)
12. Malabar anchovy - T. malabarica (Blotch)
13. Devy's anchovy - S. commersonii Lacepede
14. Devisi anchovy S. devisi (Whitley)
15. Thai anchovy S. heterolobus (Ruppell)
16. Indian anchovy S. indicus (Van Hasselt)
17. S. macrops (Hardenberg)
18. Moustached anchovy T. mystax (Scheider)
19. T. purava (Hamilton)</td>
<td>Scales and Coloration:
• Scales: Cycloid, moderate in number (about 30 to 60 in lateral series), often shed when captured.
• Coloration: Blue-green dorsum, silvery flanks, occasional silver stripe or dark saddle on the nape. Fins and mouth can vary in color, with occasional black markings but no flank spots as seen in clupeids
Four Genus.
1. Coilia:
Body tapers to a point.
Caudal fin is not forked.
2. Engraulis
No scutes before or behind pelvic fin base.
Blunt tip on the maxilla.
Anal fin origin located well behind the last dorsal fin ray.
• Engraulis encrasicolus - European anchovy
• Engraulis japonicus - Japanese anchovy
• Engraulis mordax - Californian anchovy
• Engraulis ringens - Peruvian anchoveta or Peruvian anchovy
3. Stolephorus
Scutes needle-like, found only before pelvic fin base.
4. Thryssa
Scutes present both before and behind pelvic fin base.</td>
<td>• Catch Example: In 1995, Western Central Pacific alone saw a catch of around 365,500 tons.</td>
</tr>
</table>

	20. Longjaw anchovy *T. setirostris* (Broussonet) **21. Orange mouth anchovy** *T. vitrirostris* (Gilchrist & Thompson)	**5. Setipinna** 1 st pectoral fin ray filamentous 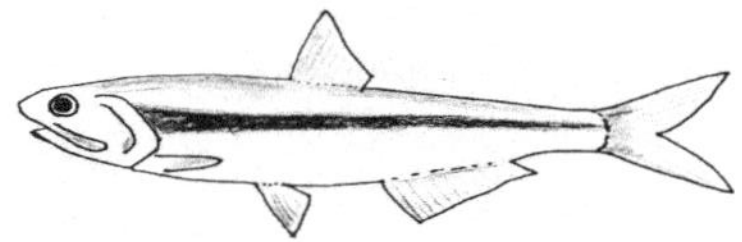	
3.	**Chirocentridae** **(Wolfherring – 1 genus and 2 species)** • **Dorab - Wolf herring** - *Chirocentrus dorab* (Forsskal) • **Whitefin - Wolf herring** *Chirocentrud nudus* Swainson	• Body elongate and compressed. Jaws with fanglike teeth (Canine). • Intestine with spiral valve. • Gill rakers 17-22. • Pyloric caeca absent. • Scales small. • Dorsal fin rays 16-18. • Pectoral fin rays 13-15. Branchiostegal rays 6-8. Cauda fin deeply forked. 	• **Distribution:** Indian Ocean (West to South Africa and the Red Sea) and western Pacific (Japan to New South Wales). • Vertebrae 70-74. Attains 3.6 m maximum length. Ravenous carnivores. Monotypic, but some authors recognize one more species.

Chapter-13

Order: Perciformes -I

Marine water fishes

Order - Perciformes

- **Habitat Preference:** Mostly Marine
- **Dorsal Fin:** Typically, Single or Two Dorsal Fins with First Spines Followed by Rays
- **Pelvic Fin:** Consists of One Spine Followed by Five Rays
- **Mouth Shape:** Conical
- **Diversity:** Extensive, Comprising 128 Families and 9293 Species

No.	Suborder	Key Characters
1.	**Percoidei**	• Maxillary bones not included in mouth gape; tooth-bearing pre-maxilla • Two or single dorsal fins with sharp spines followed by rays • Pelvic fin with single spine and rays placed at posterior part of body profile
2.	**Labroidei**	• Well-developed teeth • Parrot-like appearance in some species • Large and smooth scales • Compressed body, often colorful
3.	**Gobiodei**	• Very small fishes • Pelvic fins usually united to a cup-like structure • Gill membranes joined to isthmus • Dorsal spines present
4.	**Acanthuroidei**	• Small scales; one or two strong, movable forward-projecting spines • Protrusible premaxilla • Caudal peduncle with one spine or two bony plates
5.	**Scombroidei**	• Presence of finlets • Keels on the caudal peduncle • Large mouth; fast-swimming
6.	**Stomatoidei**	• Variable shapes • Blunt, thick snout • Small mouth; sometimes pelvic fin absent • Toothed saccular outgrowths in gullet behind last gill arch • Caudal fin with 15 branched rays

Sub-order: Percodei

S. No	Family Name	Key characters	Other features
1	**Serranidae** 1. **Redmouth grouper** - *Aethaloperca rogaa* 2. **Blue lined hind** - *Cephalopholis formosa* 3. **Tomato hind** - *Cephalopholis sonnerati* 4. **Darkfin hind** - *Cephalopholis urodeta* 5. **Coral hind** -*Cephalopholis miniata* 6. **Lapu-lapu**–*Chelidoperca pleurospilus* 7. **Aerolate grouper** – *Epinephalus areolatus* 8. **Duskytail grouper** - *Epinephalus bleekeri* 9. **Brown spotted grouper** - *Epinephalus chlorostigma* 10. **White spotted grouper** -*Epinephalus coeruleopunctatus* 11. **Orange spotted grouper** - *pinephalus coioides*	• **Operculum Structure:** Bearing Three Spines (Main Spine with One Above and One Below) • **Lateral Line:** Complete and Continuous, Excluding One Species (Lacking in One Species) • **Dorsal Fin:** Notched in Some Cases, Comprising 7-12 Spines • **Anal Fin:** Consisting of Three Spines • **Caudal Fin:** Typically, Rounded, Truncate, or Lunate; Rarely Forked • **Maxilla Position:** Tip Exposed Even with Closed Mouth • **Pelvic Fin:** One Spine with 5 Soft Rays, No Scaly Axillary Pelvic Process • **Branchiostegal Rays:** Usually 7 in Number • **Vertebrae Count:** Ranging Between 24-26 • **Reproductive Characteristics:** Monoecious with Some Functional Hermaphrodites; Protogynous Hermaphrodites in Groupers 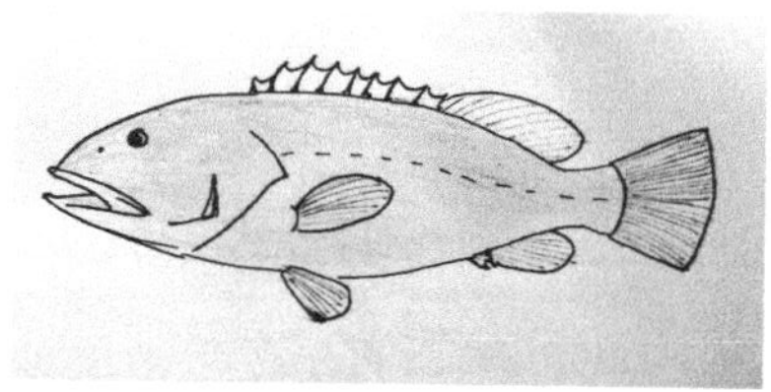	• **Distribution:** Tropical and temperate oceans. Some enter freshwater. • Anthiinae are mostly small colorful planktivores feeding primarily on tiny crustaceans and fish eggs. • They change sex from females to a few dominant males. Despite their attractive colors they need zooplankton as food and are thus not well suited for aquariums. • Groupers attain up to 3 m maximum length and weights of up to 400 kg. • They are bottom-dwelling predators and highly commercial food fish. Groupers are hardy aquarium fish, but grow rapidly. • Grammistinae get their name from a bitter tasting skin toxin, grammistin, which can kill other animals in an aquarium. They feed on crustaceans and fishes. • Subfamilies Anthiinae, Epinephelinae (tribes Epinephelini, Niphonini, Liopropomatini, Diploprioni, Grammistini) and Serraninae.

S. No	Family Name	Key characters	Other features
	12. **Spinycheek grouper** - *Epinephelus diacanthus* 13. **Cloudy grouper** - *Epinephalus erythrurus* 14. **Barredchest grouper**- *Epinephalus faveatus* 15. **Brown marbled grouper** - *Epinephalus fuscoguttatus* 16. **Giant grouper** -*Epinephalus lanceolatus* 17. **Banded grouper** - *Epinephalus latifaciatus* 18. **Longspine grouper** - *Epinephalus longispinis* 19. **Malabar grouper**- *Epinephalus malabaricus (Juv)* 20. **Dotdash grouper**- *Epinephalus poecilonotus* 21. **Oblique banded grouper** - *Epinephalus radiatus* 22. **Greasy grouper** - *Epinephalus undulosus* 23. **Blue and yellow grouper** – *Epinephalus flavocaeruleus* 24. ***Marcia's anthias*** – *Pseudanthias marcia*		• Synchronous hermaphroditism is primitive in relation to proterogyne hermaphroditism in the phylogenic evolution of Serranidae.

S. No	Family Name	Key characters	Other features
	25. **Leaflip grouper** – *Pogonoperca punctate* 26. **Peacock grouper** - *Cephalopholis argus* 27. **Chocolate hind**– *Cephalopholis boenak* 28. **Leopard hind** – *Cephalopholis leopardus* 29. **Yellow-edged lyretail** – *Vorioloa louti* 30. **White edged lyretail** – *Variola albimarginata* 31. **Double-striped soapfish** – *Diploprion bifasciatus* 32. **Camouflag grouper** – *Epinephalus polyphekadion*		

Chapter-14

Order: Perciformes -II

Marine Water Fishes

Order - Perciformes (Majority of species found in Marine water but few species Brackish water and Freshwater)

Kingdom: Animalia

Phylum: Chordata

Class: Actinopterygii (ray-finned fishes)

Order: Perciformes

Family: Silaginidae, Lactaridae, Ranchycentridae, Echeineidae, Corphaeinidae, Leiognathidae, Lutjanidae, Caesionidae, Gerridae and Mullidae.

No	Family Name	Key characters	Other features
1.	**Silaginidae** **(Smelt-Whitings - 5genus 35sp)** 1. **Northern whitning, (Silver whitning, Sand smelt)** - *(Sillago sihama* (Fabricius, 1775)	**Physical Description:** • Elongated Body Structure • Small, Sharp Mouth **Dorsal Fins:** • First Dorsal Fin Features • Spine Count and Characteristics • Second Dorsal Fin Features • Leading Spine and Soft Ray Count **Anal Fin Details:** • Structure of the Anal Fin • Leading Spine Count • Soft Ray Count 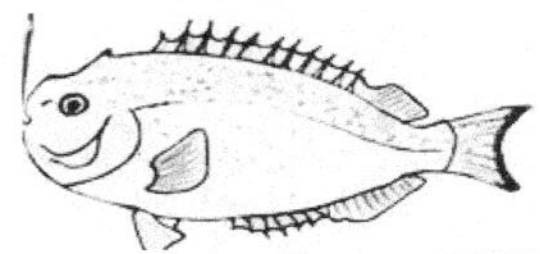	• **Distribution:** Indo-west Pacific. Silliganids are inshore species that feed mainly on benthic or epibenthic organisms. • They are esteemed table fish and some are important in estuarine aquaculture. • They attain about 45 cm maximum length.

No	Family Name	Key characters	Other features
2.	**Lactaridae** **(1 genus and 1sp)** **False travelly** - *Lactarius lactarius* (Bloch & Schneider, 1801)	**Fin structure** **Dorsal fins 2.** • Soft-rayed parts of dorsal and anal fins covered with deciduous scales (all scales easily shed). • Anal fin base is longer than that of second dorsal fin • Caudal fin forked **Skull structure** • Upper and lower jaws each with 2 small front canine teeth at front. 24 vertebrae. • Mouth large and oblique, with a prominent lower • Jaw 	• **Environment:** pelagic-neritic; brackish; marine; depth range 15 - 100 m • **Distribution:** Indo-West Pacific: East Africa to Southeast Asia, north to Japan, south to Queensland, Australia.

No	Family Name	Key characters	Other features
3.	**Echeneidae** (Remoras and Shark suckers and Disc fishes – 3genus and 8Sp) 1. **Live shark suckers** - *Echeneis naucrates* Linnaeus, 1758 2. **Whitefin shark suckers** - *Echeneis neucratoides* Zuiew, 1789 3. **Slender sucker fish** -*Phtheirichthys lineatus* (Menzies, 1791) 4. **Common Remora** - *Remora remora* (Linnaeus, 1758) 5. **Marlin sucker** - *Remora osteochir* (Cuvier, 1829) 6. **Spearfish sucker**- *Remora brachyptera* (Lowe, 1839) 7. **Whale sucker**-*Remora australis* (Bennett, 1840) 8. **White suckerfish** - *Remora albescens* (Temminck & Schlegel, 1850).	**Physical Characteristics:** • **Body Structure:** Elongate with Flattened Head • **Sucking Disc Description:** Transverse Movable Lamina • Evolution of Sucking Disc from Dorsal Fin **Mandible and Scales:** • Jutted Mandible • Small and Cycloid Scales • Absence of Spines in Dorsal and Anal Fins **Fins and Internal Features:** • Soft Ray Count in Dorsal and Anal Fins • Lack of Swim Bladder • Branchiostegal Rays Count 	Symbiotic Behavior and Feeding Habits: Relationship with Larger Hosts Host Specificity in Some Species Cleaning Behavior and Diet (Feeding on 'Leftovers') **Reproductive Characteristics:** Pelagic Spawning Behavior Assumed Spawning Method Size Range and Species **Variability:** About 1 m maximum length (reported for Echeines naucrates); the smallest species measuring 17 cm.

No	Family Name	Key characters	Other features
4	**Rachycentridae** **(Cobias – 1 genus and 1Sp)**	**Physical Structure:** • Projected Lower Jaw • **Body Shape:** Elongate with Depressed Head • **Distinctive Body Markings:** Three Dark Lateral Stripes	**Geographical Distribution:** Habitat in the Atlantic and Indo-Pacific Regions **Size and Diet:**
	1. Cobia (Black king fish, Black salmon, Ling, Lemonfish) - *Rachycentron canadum* (Linnaeus, 1766)	**Dorsal Fin Features:** • **First Dorsal Fin Description-** Short, Strong, Isolated Spines • Arrangement of Spines preceding Dorsal Fin • **Dorsal Fin Structure:** Spines and Soft Rays Count • **Characteristics of the Anal Fin:** Spines and Soft Rays Count • Absence of Membrane Connection **Caudal Fin Characteristics:** • Lunate Caudal Fin Shape in Adults • Upper Lobe Longer than Lower **Lateral Line Description:** • Wavy Pattern in the Front Part 	• **Maximum Reported Length:** Up to 1.5 meters • **Feeding Habits:** Crustaceans, Fishes, Squids • **Reproductive Behavior:** • **Assumed Spawning Method:** Pelagic Spawning • **Utilization and Significance:** Importance as Food and Gamefish

No	Family Name	Key characters	Other features
5.	**Corphaenidae** (Dolphin fishes- 1genus and 2 sp) 1. **Common Dolphin fish-** *Coryphaena hippurus Linnaeus,* 1758 2. **Pamphano Dolphin fish** - *Coryphaena equiselis Linnaeus,* 1758	**Physical Characteristics:** • Slender, Compressed Head and Body Structure • Description of the Single Dorsal Fin • Extends from Head to Nearly Full Length of Body • **Soft Ray Count:** 48-65 • Lack of Spines in Dorsal and Anal Fins • Deeply Forked Caudal Fin **Additional Notable Features:** • Beautiful Coloration in Live Specimens • Curved Lateral Line Above Pectoral Fin • Extraordinary Length of Dorsal and Anal Fins, Almost Reaching Caudal Fin **Origin of Dorsal Fin on the Nape** 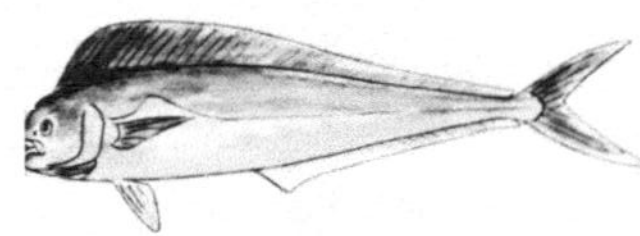	• **Geographical Distribution:** • Presence in Atlantic, Indian, and Pacific Oceans • **Distinctive Features:** • Adult Male Characteristics- Steep and High Forehead • Development of Bony Crest on Front of Head • **Anatomical Traits:** • **Vertebrae Count:** 30-34 • **Maximum Attainable Length:** Up to 1.5 meters • **Habitat and Feeding Behavior:** • **Surface Water Habitat** • **Diet:** Feeding on Small Fishes and Other Animals

No	Family Name	Key characters	Other features
6.	**Leiognathidae** (Slimys, slipmouths, or ponyfishes- 10 genus and 51sp)	**Morphological Features:** • **Body Structure:** Strongly Compressed and Slimy • Characteristics of Scales and Head • Small Scales • Naked Head with Bony Ridges on Upper Surface • United Gill Membranes with Isthmus • Small, Very Protractile Mouth • Lack of Palatal Teeth • Absence of Pseudobranchiae • **Dorsal and Anal Fin Structure:** Spines and Soft Rays Count • Locking Mechanism in Dorsal and Anal Fin Spines • Presence of Scaly Sheath at Fin Bases • **Vertebrae Count:** 22-23 • Presence of Esophageal Luminous Organs • Mucus Production (Reflecting the Common Name) 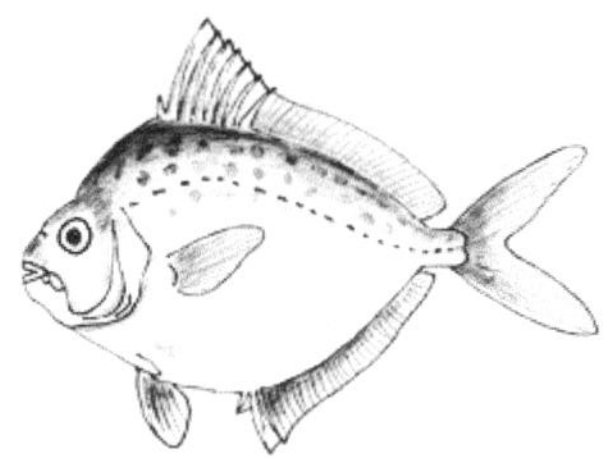	**1. Habitat and Distribution** • Marine Dominance with Some Freshwater Species • Indo-West Pacific Distribution • Occurrence in Mediterranean via Suez Canal

No	Family Name	Key characters	Other features
	1. **Barber pony fish –** *Leignathus berbis* 2. **Shortnose bony fish –** *L. brevirostris* 3. **Orangefin pony fish Photopectoralis bindus –** 4. **Gold stripe pony fish –** *L. daura* 5. **Ornate pony fish –** *L. lineolatus* 6. **Small-**toothed bony fish – *Gazza achlamys* 7. **Slender pony fish –** *Equulites elongatus* 8. **Whipfin pony fish –** *E. leucisus* 9. **Dussumier's pony fish –** *L. dussumieri.* 10. **Common pony fish –** *L. equulus.* 11. **Striped pony fish –** *L. fasciatus* 12. **Deep pugnose pony fish –** *Secutor ruconius.* 13. **Pugnose pony fish –** *Secutor insidiator.*	• **Leiognathus:** Mouth horizontal, pointing forward or downward when protracted, canniform teeth absent. • **Eg:** Leiognathus bindus, L. equilus • **Secutor :** Mouth oblique, pointing upward when protracted; canniform teeth absent • **Eg:** Secutor ruconius • **Gazza:** Mouth pointing forward when protracted; canniform teeth present. • **Eg:** Gazza achlamys	**Preferred Habitat and Feeding Behavior:** • Habitat Preference in Shallow Coastal Waters and Tidal Creeks • Feeding Habit on Benthic Invertebrates **Use and Significance:** • Catchability by Trawls or Beach Seines • Importance as an Artisanal Food Fish

No	Family Name	Key characters	Other features
7.	**Lutjanidae** (Snappers and Job fishes – 17genus and 113 sp).	**Fins and Dental Structure:** • **Description of Dorsal Fin-** Continuous or Slightly Notched, Spine Count: 10-12, Soft Rays: 10-17 • **Anal Fin Composition-**Three Spines, Soft Rays: 7-11	**Habitat and Distribution:** • Marine Habitat, Occasional Estuarine Entry
		Position of Pelvic Fins and Mouth Structure • Pelvic Fins Originating Behind Pectoral Base • Moderate to Large, Terminal Mouth • Enlarged Canine Teeth in Jaws • Small Palatine Teeth, Vomer Description • Maxilla Covered by Preorbital with Closed Mouth **Scale Distribution:** • Presence and Absence of Scales on Different Facial Areas • Absence on Anterior Head (Snout and Preorbital) • Presence on Cheek, Preopercle, and Gill Cover in Rows 	**Tropical and Subtropical Regions:** Atlantic, Indian, and Pacific Oceans • Occasional Freshwater Entry for Feeding • Suitability for Aquariums Rapid Growth Rate Limitation in Captivity **Utilization and Risks:** • Value as a Food Fish • Potential for Causing Ciguatera Poisoning **Anatomical Features:** • Branchiostegal Rays Count: 7 • Vertebrae Composition: 24 (10 + 14) • Maximum Attainable Length: Up to 1 meter • Serrated Preopercle **Feeding Behavior and Diet:** • Predatory Behavior: Predation on Crustaceans and Fishes • Some Species as Planktivores

No	Family Name	Key characters	Other features
8.	**Caesionidae** (Fusiliers- 4genus and 23 sp)	**Morphological Characteristics:** • **Body Shape:** Oblong to Fusiform • **Dorsal Fin:** Spine and Soft Ray Counts • **Caudal Fin Structure:** Distinctly Forked with Pointed Lobes • **Scale Distribution:** Horizontal Rows on Body, Exceptions in Gymnocaesio gymnoptera and Dipterygonotus balteatus **Mouth and Dentition:** • Small Highly Protrusible Mouth • **Description of Teeth:** Small or Minute Conical **Coloration and Markings:** • Axil of Pectoral Fins-Black Marking • Variations in Side Coloration and Caudal Fin Markings 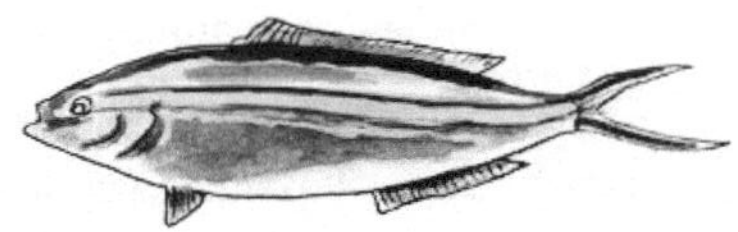	• **Geographical Distribution:** • Indo-West Pacific Distribution • **Anatomical Features:** • **Description of Head Structure:** Longitudinal Axis and Dentition Details • **Branchiostegal Rays Count**: 7 • **Scale Size and Type:** Moderate to Small, Weakly Ctenoid • **Lateral-Line Scales Count:** Range from 45 to 88 • **Adaptations and Trophic Ecology:** • Adaptations for Planktivorous Mode of Life • **Feeding Behavior:** Mid-water Zooplankton Feeding Schools, Cleaning Stations • Behavior During Day and Night. • **Ecological Suitability and Use:** • Habitat Preferences and Space Requirements • Unsuitability for Home Aquariums • Importance as Food Fish and Bait in Tuna Fisheries

No	Family Name	Key characters	Other features
9.	**Gerridae** **(Silver biddies and Mojarras- 8 genera and 53 sp)** **1. Silver biddy** – *Gerres filamentous.*	**Morphological Characteristics:** • Highly Protractile Mouth • Scaly Head Structure with Smooth Upper Surface • Presence of Scaled Sheath Along the Base of Dorsal and Anal Fins • Gill Membranes Not United to Isthmus • **Description of Tail:** Deeply Forked • **Vertebrae Count:** 24 • **Maximum Attainable Length:** Up to 35 cm in Gerres filamentosus 	**Geographical Distribution:** • Presence in Most Tropical Seas • Primarily Marine Habitat with Occasional Brackish Water and Rare Freshwater Occurrence **Feeding Behavior:** • **Description of Feeding Method:** Sorting Benthic Invertebrates from Sand **Ecological Role and Habitat:** • Importance as Food Fishes **Assumed Reproductive Behavior:** • Non-Guarders • Adult Habitat • **Preference:** Coastal Lagoons with Sandy or Muddy Bottoms, Often Bordered by Mangroves • Occasional Entry into River Mouths **Juvenile Behavior:** • Estuarine Entry until Maturity **Spawning and Reproduction:** • Spawning Occurrence at Sea Throughout the Seasons

No	Family Name	Key characters	Other features
10.	**Mullidae** **(Goat fishes- 6 genera and 100 sp)**	**Physical Characteristics:** • **Body Structure:** Elongated • **Arrangement of Dorsal Fins:** Far Apart • **First Dorsal Fin:** 6-8 Spines • **Second Dorsal Fin:** One Spine, 8-9 Soft Rays (Shorter than Anal Fin) • **Anal Fin Composition:** 1-2 Spines, 5-8 Soft Rays • **Caudal Fin Shape:** Forked • **Vertebrae Count:** 24 **Distinctive Features:** • **Chin Description:** Two Long Barbels • **Functions:** Contain Chemosensory Organs, Used for Probing Sand or Reef Holes for Prey • Bright Coloration in Many Species 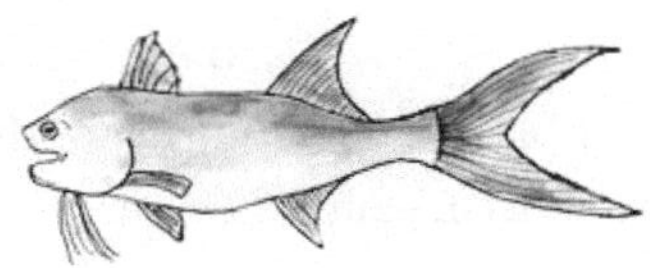	**Geographical Distribution:** • Presence in Atlantic, Indian, and Pacific Oceans • Occasional Occurrence in Brackish Waters **Reproductive Behavior:** • **Spawning Behavior:** Pelagic Spawners • **Size and Utilization:** • **Maximum Length:** Up to 60 cm • Importance as a Food Fish • **Suitability for Aquariums:** Not Popular but Successful in Specific Settings **Habitat and Feeding Behavior:** • **Preferr,ed Habitat:** Sand-associated Shallow • Habitats • **Feeding Habits:** Hunting • Benthic Invertebrates or Small Fish Using Barbels

Chapter-15

Order: Perciformes -III

Marine Water Fishes

Order - Perciformes (Majority of species found in Marine water but few species Brackish water and Freshwater)

Kingdom: Animalia

Phylum: Chordata

Class: Actinopterygii (ray-finned fishes)

Order: Perciformes

Suborder: Percodei

Family: Haemulidae, Sparidae, Lethrinidae, Nemipteridae, Polynemidae, Scianidae, Mullidae, Carangidae, Monodactylidae, Chaetodontidae, Drepanidae, Pomacanthidae, Teraponidae.

No	Family Name	Key characters	Other features
1.	**Haemulidae** (Grunts, Swesetlips, Rubber lips and Hotlips) **Striped sweetlips** – *Plectorinchus diagrammus.* **Yellow banded sweetlips** – *P. lineatus.* **Minstrel sweetlips-** *P. schotaf.* **Indian ocean sweetlips-** *P. vittatus.* **Giant sweetlips-** *P. albovittatus.* **Trout thicklips-** *P. pictus.* **Indian grunter –** *Pomadysus guoraca.* **Javelin grunter-** *P. kaakan.* **Saddle grunt-** *P. maculatum.* **Bluecheek silver grunt –** *P. argyreus.* **Banded grunter-** *P. furcatus.* **Painted sweetlips-** *Diagramma pictum* **Srilankan sweetlips-** *Plectorhichus ceylonensis.*	**Fin structure:** • Dorsal fin continuous, 9-14 spines, 11-26 soft rays. • Anal fin 3 spines, 6-18 soft rays. Small mouth with thick lips. • Usually cardiform jaw teeth, with vomer generally toothless. • Usually with enlarged chin pores. • Branchiostegal rays 7. Vertebrae 26 or 27 (10 or 11 + 16). **Skull structure:** • Scales present on entire head (Except front of snout, tips and chin. • Chin with 2 pores anteriorly and a median pit or 6 pores and no pit in Ptectorhincus • Preopercle with a slightly concave and serrated posterior margin. • Tips thick, tip of upper jaw hidden when mouth closed. 	• **Distribution:** Atlantic, Indian and Pacific. Chiefly marine, some brackish, rarely freshwater. • Adults are typically inactive during day when they shelter near or under ledges; they disperse to feed on benthic invertebrates at night. • Distinct pairing during breeding producing pelagic eggs with no known parental care (Ref. 205). Pelagic spawners. • Juveniles are popular in-home aquaria, but adults require a large tank. Maximum length about 60 cm. • Important food fishes. Includes the former Isnermiidae.,

No	Family Name	Key characters	Other features
2.	**Sparidae** (Porgies, Seabreams, Stumpnoses)	• Head large, often with a steep upper profile, cheeks scaly, preopercle with or without scales, without spines or serrations on margin • Dorsal fin single, with 10 to 13 spines and 9 to 17 soft rays; no notch between spiny and soft portions • Anal fin with 3 spines and 7 to 15 soft rays • An axillary scale at the base of pelvic fins 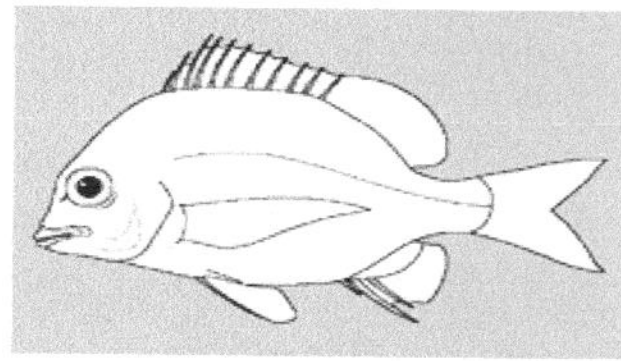	• **Distribution:** tropical and temperate Atlantic, Indian and Pacific Oceans. Habitat: chiefly marine; very rare in fresh- and brackish water. • Usually most common along the shore from shallow water (including estuaries), to deeper water as demersal inhabitants of the continental shelf and slope. • **Trophic**: most are carnivorous, feeding on benthic invertebrates. • **Carnivores of hard**-shelled benthic invertebrates. Many species have been found to be hermaphroditic; some have male and female gonads simultaneously; others change sex as they get larger. • **Uses:** Premier food and game fishes. Many species around southern Africa. A few species have been implicated in cases of ciguatera. • 6 Subfamilies (defined by tooth morpholgy).

No	Family Name	Key characters	Other features
3.	**Lethrinidae** (Emperors, Pig face bream, Large eye breams - 5 Genus and 42 Species) 1. **Gray largeeye-** *Gymnocranius griseus.* 2. **Pinkear emperor-** *Lethrinus lentjan* 3. **Thumbprint emperor-** *L. harak* 4. **Smalltooth emperor-** *L. microdon.* 5. **Gold-lined seabream-** *L. nebuloses.* 6. **Longface emperor-** *L. olivaceaus.* 7. **Ornate emperor-** L. ornatus.	• Short gill rakers that are knob-like. • The dorsal fin is single and continuous, with 10 spines and 9 or 10 branched (soft) rays. • Cheeks, upper surface of the head, and preorbital area are scaleless in the genus Lethrinus, while other genera in Lethrinidae have scales on the cheek. • The maxilla overlaps the premaxilla. 	• **Lethrinidae,** specifically the Lethrinus genus within that family. These fish are found in tropical waters of the Indo-Pacific, with one species, Lethrinus atlanticus, occurring in the Atlantic off West Africa. • Lethrinids are bottom-feeding, carnivorous, coastal fishes, ranging primarily on or near reefs. They typically feed primarily at night on benthic invertebrates or fishes, those with molariform teeth mainly on hard-shelled invertebrates. • Lethrinids can be solitary or schooling and do not appear to be territorial. They often form large aggregations for pelagic spawning. • Protogynous hermaphroditism (sex reversal from female to male) has been demonstrated in several species of Lethrinus.

No	Family Name	Key characters	Other features
4.	**Nemipteridae** (Thread fin breams / Monocle breams / Dwarf monocle breams-5 genus and 73 sp). **1. Delagoa threadfin bream-** *Nemipterus bipunctatus.* **2. Fork-tailed threadfin bream-** *N. furcosus.* **3. Japanese threadfin bream-** *N. japonicus.* **4. Randall's threadfin bream-** *N. randalli.* **5. Rosy dwarf monocle bream-** *Parascolopsis eriomma.* **6. Yellow-striped monocle bream-** *Scolopsis auratus.* **7. Thumbprint monocle bream-** *S. bimaculatus.* **8. Whiteneck monocle bream-** *S. vosomeri.* **9. Oblique- barred monocle bream-** *S. xenochrous.*	• Single dorsal fin originating above pectoral fin • bases, the first spine sometimes prolonged into a filament • Just above pelvic fin, medium sized axillary scale present • Caudal fin forked or emarginated, upper lobe sometimes with a filament • Dorsal fin with 10 spines; soft rays 9. • Three spines in anal fin; soft rays 7 (Nemipterus virgatus with 8). • A filament off upper lobe of caudal fin in some species. • Well developed subocular shelf and accessory pelvic keel. • Intercalar present. • A backward pointing spine below eye in Scolopsis and in some Parascolopsis species • Front of head scaleless, scales on top of head (predorsal scales) beginning above or just in front of Eye. • Cheeks with 3 rows of scales in Nemipterus, 4 to 7 rows in Scolopsis and Parascolopsis 	• Distribution: Tropical and sub-tropical Indo-West Pacific. • Carnivorous fishes that feed mainly on benthic small fishes, cephalopods, crustaceans and polychaetes; some species are planktivores. • Protogynous hermaphroditism is reported for some species of Scolopsis. Not generally kept in aquaria, but potentially hardy. • Threadfin breams are an important component of commercial and artisanal fisheries.

No	Family Name	Key characters	Other features
5.	**Polynemidae** *(Thread fins/Tassel fishes – 8 genus and 42 sp)* 1. **Fourfinger threadfin-** *Eleutheronema tetradactylum* 2. **Indian sevenfinger threadfin-** *Fillimanus similis* 3. **Striped threadfin-** *4. Polydactylus plebius* 5. **Sixfinger threadfin-** *P. sexfilis* 6. **Blackspot threadfin-** *P. sextarius.*	• Mouth inferior. Pectoral fin with 2 sections. • Upper section of pectoral fin with the rays attached; the lower section having 3-7 (except in Polistonemus with 14 or 15) long free rays. • Spinous and soft dorsal fins far apart. Pelvics subabdominal. • One spine in pelvic fin; soft rays 5. Deeply forked caudal fin. • Lateral line extending upto caudal fin. 	• Chiefly marine and brackish. Some riverine. Distribution: all tropical and subtropical seas. • Vertebrae 24 or 25. Reaches 1.8 m maximum length, reported for Eleutheronema tetradactylum. Feed on benthic invertebrates of sandy to muddy bottoms. Few species near reefs. Foodfishes. Assumed to be pelagic spawners

No	Family Name	Key characters	Other features
6.	**Sciaenidae** (Croakers, Drums, Meagres – 69 genus and 297 sp). **1. Caroun croaker-** *Jhonius carouna* **2. Tiger-toothed croaker-** *Otolithes ruber* **3. Panna croaker-** *Panna microdon* **4. Karut croaker-** *Johnius carutta* **5. Kathala croaker-** *Kathala axillaris.* **6. Bloched croaker-** *Nibea maculate* **7. Greyfin croaker/Donkey croaker-** *Pennahia anea* **8. Black-spotted croaker-** *Protonibea diacanthus* **9. Blotched tiger-thoothed croaker-***P. maculatus.*	**1. Scaled Head and Body:** • Except at the extreme part of the snout, both head and body are usually scaled. **2. Sensory Pores:** • Present on various parts of the snout - rostral pores on the tip of the snout, marginal pores on the lower edge of the snout, and mental pores on the chin. • Typically, there are 3 to 5 rostral pores on the top of the snout, 5 nearer the anterior margin of the mouth, and 3 pairs on the lower jaw. **3. Feeding Habits and Pore Development:** • Bottom feeders such as some species of Johnius have well-developed rostral and mental pores. • Mid-water feeders like Otolithes and Argyrosomus tend to have indistinct pores. **4. Mouth Position:** • Terminal mouth in species like Otolithes and Pennalia. • Subterminal in Protonibea. • Inferior or lower jaw projecting in Johnius and Umbrina. **5. Fin Features:** • Dorsal fin is long and continuous, often with a deep notch between the anterior (spinous) and posterior (soft) portions. • Anal fin typically has 2 spines, with the second spine enlarged.	• Drums or croakers are distributed in the Atlantic, Indian and Pacific oceans. • Drums or croakers are distributed in the Atlantic, Indian and Pacific oceans. • Large cavernous canals in head. Snout and lower jaw with conspicuous pores. Vomer and palatine toothless. • Swim bladder usually having many branches and used as a resonating chamber. • Exceptionally large otoliths. Vertebrae 24-29. Bottom dwelling carnivores, feeding on benthic invertebrates and small fishes. • Juveniles are popular aquarium fishes, but difficult to maintain. • Eggs are spherical, transparent, and pelagic

No	Family Name	Key characters	Other features
		6. Lateral Line and Scales: • The lateral line extends to the hind margin of the caudal fin. • Scales may vary, with cycloid scales on the head/ ctenoid scales on the body, or only cycloid scales in certain species like Jhonius dussumieri. **7. Internal Characteristics for Identification:** • Internal features such as the shape and size of otoliths (ear bones) and swim bladder can be helpful in identifying genera/species.	
7.	**Carangidae** (Jacks / Trevallies / Scads / Queenfishes / Amberjacks / Pilotfishes / Pampanos / Leerfish – 31 genus and 138 sp). • **Indian threadfin fish-** *Alectis indicus.* • **African pompano-** *A. indicus.* • **Shrimp scad-** *Alepes djedaba.* • **Razorbelly scad-** *Alepes kleinii.* • **Herring scad-** *A. vari.* • **Cleftbelly trevally-** *Atropus atropus.* • **Yellowtail scad-** *Atule mate.* • **Bigeye trevally-** *Caranx sexfasciatus.* • **Blacktip trevally-** *Caranx heberi.* • **Giant trevally-** *Caranx ignobilis.*	**1. Dorsal Fins:** • In small juveniles, these fish have two separate dorsal fins. • The first dorsal fin is moderate in height or very low. • The anterior lobe of the second dorsal fin can range from scarcely produced to extremely long. **2. Anal Fin:** • The anal fin typically has two anterior spines (detached spines). • The anterior lobe of the anal fin is described as low to elongate in shape.	• Chiefly marine; rarely brackish. Atlantic, Indian and Pacific Oceans. • Fast swimming predators of the waters above the reef and in the open sea. Some root in sand for invertebrates and fishes. • One of the most important families of tropical marine fishes; fished commercially and for recreation. ,

No	Family Name	Key characters	Other features
	• **Orange-spotted trevally-** *Carangoides bajad.* • **Longnose trevally-** *C. chrysophrys.* • **Coastal trevally-** *C. caeruleopinnatus.* • **Bumpnose trevally-** *C. hedlandensis* • **Coachwhip trevally-** *C. oblongus.* • **Brownback trevally-** *C. praeustus.* • **Indian pompano-** *Trachinotus mookalee.* • **Mackeral scad-** *Decapterus macarellus.* • **Indian scad-** *D. russelli.* • **Rainbow runner-** *Elagatis bipinnulata.* • **Smallspotted dart-** *Trachinotus baillonii.* • **Golden trevally-** *Gnathanodon speciosus.* • **Whitetongue jack-** *Urapis helvola.* • **Torpedo scad-** *Megalaspis cordyla* • **Black pomfret-** *Parastromateus niger* • **Talang queenfih-** *Scomberoides commersonianus.*	• **3. Scutes in Lateral Line:** • Scutes, which are enlarged, thickened, and often pointed scales in the lateral line, are a prominent feature in these fish. • However, in some species, these scutes may be reduced. • In certain genera, they might be entirely absent. 	

No	Family Name	Key characters	Other features
	• **Double-spotted queen fish-** S. lysan • **Needle-scaled queen fish-** *S. tol.* • **Big eye scad-** *Salar crumenophthalmus.* • **Yellowstripe scad-** *Selaroides leptolepis.* • **Blackbanded trevally-** *Seriolina nigrofasciata* • **Snubnose pompano-** *Trachinotus blochii.*		
8.	**Monodactylidae** (Moonyfishes or fingerfishes – 2 genus and 6sp) • **Sliver moony-** *Monodactylus argenteus.* • **Kottelat's moony-** *M. kottelati.*	• Body deep and highly compressed. • Pelvic fins present in juveniles, lacking or vestigial in adults in Monodactylus. • Dorsal fin with the base long and scaly; 5-8 short and graduated spines. • Anal fin base long; 3 spines. Scales cycloid or ctenoid. Often silvery. 	• Distribution: west Africa, Indo-Pacific. Chiefly marine and brackish; occasionally entering freshwater.

No	Family Name	Key characters	Other features
9.	**Cheatodontidae** (Butterfly fishes- 12 genus and 136 sp) 1. **Threadfin butterfly fish-** *Chetodon Auriga.* 2. **Redtail butterfly fish-** *C. collare.* 3. **Blackback butterfly fish-** *C. melannotus.* 4. **Black-wedged butterfly fish-** *Chaetodon falcula.* 5. **Gardiner's butterfly fish-** *C. gardineri.* 6. **Lined butterfly fish-** *C. lineolatus.* 7. **Indian vagabond butterfly fish-** *C. decussatus.* 8. **Eightband butterfly fish-** *C. octofasciatus.* 9. **Chervron butterfly fish-** *C. trifacialis.* 10. **yellowhead butterfly fish-** *C. xanthocephalus.* 11. **Longfin banner fish-** *Heniochus acuminatus.* 12. **Phantom bannerfish-** *H. pleurotaenia.*	**1. Body Characteristics:** • Highly compressed body shape. • Typically deep and strongly compressed body structure. • Snout can range from slightly produced to greatly elongated. **2. Fin Structure:** • Dorsal fin is continuous, consisting of 6-16 spines and 15-30 soft rays. • Anal fin has 3-5 (usually 3) spines and 14-23 soft rays. • Caudal fin possesses 15 branched rays and is rounded to emarginate in shape. • Scales extend onto both the anal and dorsal fins. **3. Mouth and Teeth:** • Small, terminal, and protrusible mouth. • Often characterized by a band or rows of small brushlike teeth. **4. Internal Characteristics:** • Gut typically coiled several times. • Swim bladder with two anteriorly directed processes. • A total of 24 vertebrae, organized as 11+13. **5. Coloration and Markings:** • Most species exhibit bright coloration. • A dark band across the eye is a common feature. • Presence of an 'eyespot' dorsally is typical.	• Atlantic (tropical to temperate), Indian, and Pacific Oceans. Primarily Indo-west Pacific. • Generally near coral reefs. Typically diurnal. Many feed on a combination of coelenterate polyps or tentacles, small invertebrates, fish eggs, and filamentous algae while others are specialists or planktivores. • Most species occur as heterosexual pairs. Pelagic spawners. • Tholichthys larval stage with the head region covered with bony plates. • Some of the planktivores and generalists do well in the aquarium, but most species are difficult to maintain, and obligate corallivores nearly impossible.

No	Family Name	Key characters	Other features
		6. Additional Features: • Axillary scaly process present at the base of the pelvic fin spine. • Dorsal fin is continuous, sometimes with a slight notch between the spinous and soft portions. • Possession of ctenoid scales. 	
10.	**Drepanidae** (Sickle fishes – 1 genus and 3 species) **1. Spotted sickle fish** - *Drepane punctata (Linnaeus, 1758).* **2. Barred sickle fish** - *Drepane longimana (Bloch & Schneider, 1801).* **3. African sickle fish** - *Drepane africana Osório, 1892.*	• Body deep and laterally compressed. Mouth markedly protractile. • Spinal portion of dorsal fin with 13-14 spines, distinct from soft-rayed portion with 19-22 soft rays; anal fin with three spines and 17-19 soft rays; pectoral fins longer than head, falcate; 24 vertebrae. • Maxilla distally exposed. Subocular shelf absent. 	• Indo-west Pacific and West Africa. • Feed on small invertebrates. Assumed to be pelagic spawners. • Drepane punctata and D. longimana are only distinguished by color and might be the same species. • Name was changed from Drepanidae (preoccupied in insects) to Drepaneidae,

No	Family Name	Key characters	Other features
11.	**Pomacanthidae** (Angelfishes – 8 genus and 91sp) 1. **Dusky angel fish/Brown dwarf angel fish-** *Centropyge multispiniss* 2. **Yellowtail angel fish/ Indian smoke angel fish-** *Apolemichthys xanthurus* 3. **Pomacanthus imperator-** *Emperor angel fish.* 4. **Koran angel fish/ Semicircle angel fish-** *P.semicirculatus.* 5. **Bluering angel fish-** *P. annularis.*	**Elongate body -** oval to orbicular in shape/strongly compressed • Mouth protrusible • Dorsal fin single, continuous without a notch • A strong spine at an angle in preopercle. 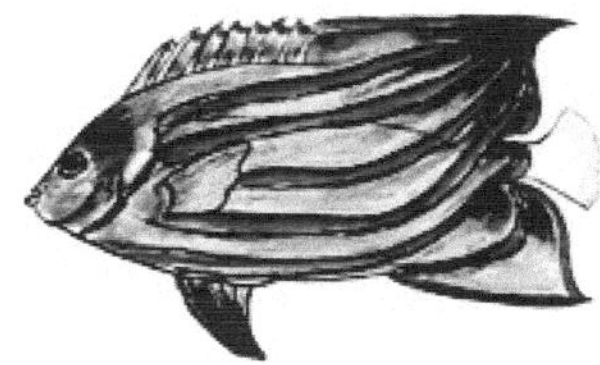	• Tropical Atlantic, Indian, and (mainly western) Pacific. • Striking coloration, markedly different between juveniles and adults of many species. In shallow waters of less than 20 m deep, very seldom below 50 m; generally near coral reefs. • Social organizations in this group vary from monogamy, harems, promiscousity, and lekking. • Several species are protogynous hermaphrodites engaging in 'haremic' social system. Pelagic spawners. Species of Centropyge feed primarily on filamentous algae and species of Genicanthus feed primarily on zooplankton; most others feed on sponges, invertebrates, algae and fish eggs. • Most species do well in the aquarium, but some food specialists are difficult to maintain. Poma- = operculum, acanth = spine. • Three basic feeding styles: predation on sessile invertebrates, herbivory and planktivory.

No	Family Name	Key characters	Other features
12.	**Teraponidae** (Terapon-perches / Terapons). Terapon jarbua- Crescent Terapon T.puta- Small scaled terapon T. theraps- Largescaled terapon.	Preopercle serrate/serrations more prominent in juveniles Two opercular spines; lower one larger and stronger; post temporal bone exposed in some species, its posterior margin serrate Single dorsal fin; with 11 to 14 spines Anal fin with 3 strong spines. 	• Terapons inhabit inshore marine and brackish waters and some species enter freshwaters; • most marine species with 3 or more longitudinal body stripes. • bout 80 cm maximum length. Feed on fishes, insects, algae, and sand-dwelling invertebrates. • Also spelled Teraponidae, Theraponidae or Therapontidae.

Chapter-16

Order: Perciformes -IV

Marine Water Fishes

Order - Perciformes (Majority of species found in Marine water but few species Brackish water and Freshwater)

Kingdom: Animalia

Phylum: Chordata

Class: Actinopterygii (ray-finned fishes)

Order: Perciformes

Suborder: Labrodei

Family: Pomacentridae, Labridae and Scaridae.

No	Family Name	Key characters	Other features
13.	**Pomacentridae** (Damsel fishes - 30 genus and 419 sp). 1. **Bengal sergeant** – *Abudefduf bengalensis* 2. **Indo-Pacific sergeant** – *A. vaigiensis* 3. **Banded sergeant**- *A. septemfasciatus.* 4. **Yellowtail clown fish**- *Amphiprion clarkia* 5. **Sebae anemone fish** – *A. sebae* 6. **Gulf damsel fish**- *Pristotis obtusirostris.* 7. **Raticulate damsel**- *Dascyllus reticulatus.* 8. **Threespot damsel**- *D. trimaculatus.* 9. **Threestripe damsel**- *D. aruanus.* 10. **Green chromis** – *Chromis viridis* 11. **Coral demoiselle**- *Neopomacentrus nemurus* 12. **Caerulean damsel/ blue yellow damsel** – *Pomacentrus caeruleus.*	• **Nostrils:** Typically, these fish have one nostril on each side of their head. However, some species within the genera Chromis and Dascyllus might possess double nostrils. • **Body Shape:** Damselfishes usually have a deep and compressed body structure, which aids in their maneuverability within reefs and rocky areas. • **Mouth Size:** They typically have small mouths, adapted for feeding on small invertebrates and algae. • **Lateral Line:** Their lateral line is often incomplete and interrupted. • **Anal Fin:** The anal fin usually contains 2 spines, rarely 3. • **Teeth:** Damselfishes generally lack palatine teeth. • **Size:** Maximum length for these fish species is around 35 centimeters. • **Coloration:** The coloration of damselfishes can vary significantly among individuals and can also be influenced by the environment or locality where they reside. 	• **Environment:** Chiefly marine; rare in brackish water and freshwater; common in coral reefs Range: All tropical seas, mainly Indo-Pacific; • Many species are highly territorial herbivores, omnivores, or planktivores. • Damselfishes lay elliptical demersal eggs that are guarded by the males. Included are the anemonefishes (Amphiprioninae), which live in close association with large sea anemones. • Damsels are among the hardiest aquariumfishes, but some species are extremely aggressive.

No	Family Name	Key characters	Other features
14.	**Labridae** (Wrasses, Hogfishes, Razor fishes, Coris, Tuskfishes – 70 Genera and 562 sp) 1. **Bay of Bengal hog fish –** *Bodianus neilli.* 2. **Floral wrasse-** *Cheilinus chlorurus.s* 3. **Blackeye thicklip-** *Hemigymnus melapterus* 4. **Tripletail wrasse-** *Chelinus trilobatus.* 5. **Humphead wrasse-** *Cheilinus undulates.* 6. **Pale barred coris-** *Coris dorsomacula.* 7. **Green bird mouth wrasse-** *Gomphosus caeruleus* 8. **Checkerboard wrasse-** *Halichoeres hortulanus* 9. **Dusky wrasse-** *Halichoeres marginatus* 10. **Bubblefin wrasse-** *H. nigrescens* 11. **Goldstripe wrasse-** *H. zeylonicus* 12. **Bluestreak cleaner wrasse-** *Labroides dimidiatus.* 13. **Cocktail wrasse-** *Pteragogus flagelifer* 14. **Cutribbon wrasse-** *Stethojulis interrupta* 15. **Sixbar wrasse-** *Thalassoma hardwicki*	**Dorsal and Anal Fin Structure:** The dorsal fin typically consists of 8-21 spines (often fewer than 15) and 6-21 soft rays. The anal fin generally has 4-6 spines (sometimes 3) and 7-18 soft rays. **Scales and Vertebrae:** They possess cycloid scales, which are generally moderate to large in size, ranging from 25 to 80 along the side. The number of vertebrae usually falls between 23 and 42. **Lateral Line:** The lateral line can be interrupted or continuous. It may have a smooth curve or an abrupt curve below the soft portion of the dorsal fin, either being continuous or interrupted below the posterior part of the dorsal fin. **Snout Variation:** The genus Gomphosus within this family tends to have an elongated snout. **Size, Shape, and Color:** Wrasses exhibit extensive variation in size, body shape, and coloration. This diversity is often noticeable within the same species across different stages of their life cycle or based on gender differences. **Mouth Structure:** Their mouths are terminal, with prominent lips, suited for their varied feeding habits. 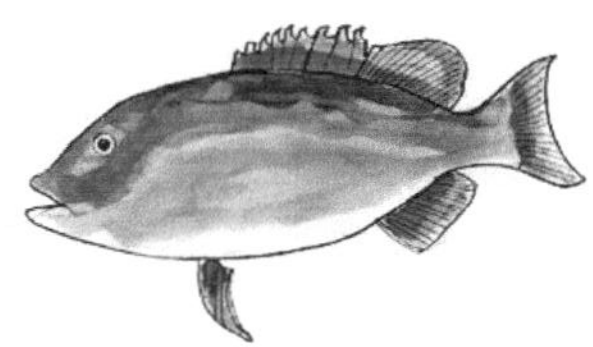	• **Distribution:** Atlantic, Indian, and Pacific. Protrusible mouth. Most jaw teeth with gaps between them; teeth usually jutting outward. • Most species are sand burrowers; carnivores on benthic invertebrates; also planktivores, and some small species remove ectoparasites of larger fishes. • Most species change color and sex with growth, from an initial phase (IP) of both males and females, the latter able to change sex into an often brilliantly colored terminal male phase (TP). • Males dominate several females; all Indo-Pacific species are pelagic spawners. Most species do well in aquaria, and young Coris are particularly popular. Maximum length about 2.3 m, many are less than 15 cm, the shortest being 4.5 cm.

No	Family Name	Key characters	Other features
	16. **Moon wrasse-** *T. lunare* 17. **Jaansen's wrasse-** *T. jansenii* 18. **Twospot razor fish-** *Iniistius bimaculatus.* 19. **Peacock wrasse-** *Iniistius pavo* 20. **Robust dusk fish-** *Choerodon robustus.* 21. **Blue banded wrasse-** *Xiphochelius typus.* 22. **Cigar wrasse-** *Cheilio inermis* 23. **Queen coris-** *Coris formosa*		• Medium to large species are important food fishes.
15.	**Scaridae** (Parrot fishes- 10 genus and 10 sp.) 1. **Knothead parrot fish-** *Chlorurus oedema* 2. **Rivulated Parrot fish-** *Scarus rivulatus* 3. **Ember parrot fish-** *Scarus rubroviolaceus* 4. **Eclipse parrot fish** – *S. russelii.* 5. **Blue barred parrot fish-** *S. ghobban* 6. **Quoy's parrot fish-** *S. quoyi.* 7. **Marbled parrot fish-** *Leptoscarus vaigiensis.* 8. **Heavybeak parrot fish** – *Chlorurus gibbus.*	• **Head Shape:** Parrotfishes also have a bluntly rounded anterior head structure, similar to triggerfish. • **Teeth Structure:** Their teeth are typically fused together, forming a pair of beak-like plates in each jaw. Some species might have incisor/canine teeth as well, especially during their juvenile or initial stages. • **Fins:** Parrotfishes possess a single continuous dorsal fin, consisting of 9 slender and flexible spines followed by 10 soft rays. The anal fin usually contains 3 flexible spines and 9 soft rays	**Distribution:** Atlantic, Indian, and Pacific Oceans. . • Herbivorous, usually scraping algae from dead coral substrates. • Bits of rock eaten with the algae are crushed into sand and ground with the algae to aid in digestion, making parrotfish some of the most important producers of sand on coral reefs. At night, some species rest enveloped in their mucoid secretion.

No	Family Name	Key characters	Other features
		• . • **Caudal Fin:** Their caudal fin tends to be rounded to lunate in shape, often with produced caudal lobes. • **Scales:** Parrotfishes may have 1-4 rows of scales on the cheek area and typically 2-8 median predorsal scales. 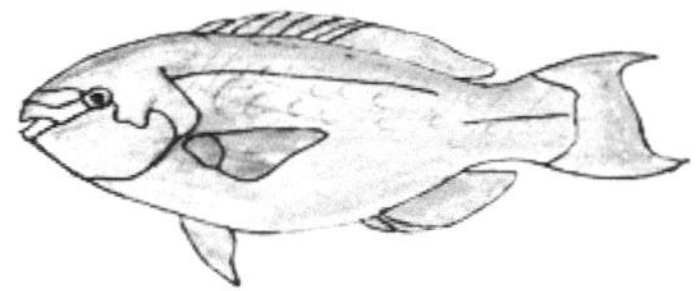	• Sex change seems a common occurrence, with an initial phase (IP) of both males and females, and the latter changing into a brilliantly colored male terminal phase (TP). Terminal males dominate several females; pelagic spawners. • Many species could be identified by their live coloration but this may be lost in preservation, or can vary between juveniles and adults and with sex change. Important food fishes. • Difficult to maintain in aquaria as the fused teeth need to constantly graze dead coral rock in order to keep from growing too long.

Chapter-17

Order: Perciformes -V

Marine Water Fishes

Order- Perciformes (Majority of species found in Marine water but few species Brackish water and Freshwater)

Kingdom: Animalia

Phylum: Chordata

Class: Actinopterygii (ray-finned fishes)

Order: Perciformes

Suborder: Gobiodei

Family: Gobidae, Mugilidae,.

No	Family Name	Key characters	Other features
1.	**Gobidae** (248 genus and 1902 sp). 1. **Komochaijacko –** *Amblychaeturichthys sciistius* 2. **Giant prawn goby-** *A. fontanesii.* 3. **Indo-pacific burrowing goby-** *Trypauchen vagina* 4. **Karamelliraputokko-** *Cryptocentrus leonis* 5. **Tail-eyed goby-** *Parachaeturichthys polynema.* 6. **Jester goby-** *Oxyurichthys paulae.*	• **Morphology:** Pelvic fins fused into an adhesive disc, when well developed. Spinous dorsal present or absent; when present with 2-8 flexible spines and discontinuous with soft dorsal. Cycloid or ctenoid scales almost always present. The pelvic fins are united and forming an adhesive or sucking dics. • First dorsal spinous and seperated from second. Five branchiostegal rays. • They live in close association with other animals like sponges and sea urchins This family contains world smallest fishes. • The marine goby Trimmatomnanus female measures 8-10 mm. The freshwater counter part • Pandaka pygmaea and Mistichthys luzonensis measure upto 10-11mm. It is most species rich family of vetebrates. 	• **Distribution:** mostly tropical and subtropical areas. Environment: Chiefly marine and brackish, some species are catadromous. • Prominent head barbels present in some species. To 50 cm maximum length; most species below 10 cm. • The largest family of marine fishes (possibly > 2,000). The smallest fishes (and vertebrates) in the world belong to this family. • Habitat and Behavior: Mostly marine in shallow coastal waters and around coral reefs. Most are cryptic bottom dwelling carnivores of small benthic invertebrates; others are planktivores. • Some species have symbiotic relationships with invertebrates (e.g. shrimps) and others are known to remove ecto-parasites from other fishes. Typically nest spawners with non-spherical eggs guarded by the male.

No	Family Name	Key characters	Other features
			• Many are popular aquarium fishes. The following subfamilies are recognized: Amblyopinae (eel gobies or worm gobies):typically inshore, mud-dwelling fishes of the Indo-west Pacific region; with elongate bodies with a continuous dorsal fn

Chapter-18

Order: Perciformes -VI

Marine Water Fishes

Order - Perciformes (Majority of species found in Marine water but few species Brackish water and Freshwater)

Kingdom: Animalia

Phylum: Chordata

Class: Actinopterygii (ray-finned fishes)

Order: Perciformes

Suborder: Acanthurodei

Family: Ephippidae, Platicidae, Acanthuridae, Scatophagidae, Siganidae and Zanciliade.

No	Family Name	Key characters	Other features
1.	**Ephippidae** (Spade fishes-8 genus and 15sp) 1. **Orbfish-***Ephippus orbis*	• Head short, its profile either steep or fairly convex • Pectoral fins rounded, small; pelvic fins thoracic, pointed with an axillary process • Anal fin with 3 spines. 	Atlantic, Indian, and Pacific; marine, rarely brackish. Omnivores of algae and small invertebrates. Thought to be pelagic spawners. Juveniles of Platax species are popular and unproblematic aquarium species, but grow very fast.
2.	**Platicidae** (Bat fishes) 1. **Orbicular bat fish-** *Platax orbicularis* 2. **Longfin bat fish-** *P. teira.*	• Deep body, rhomboid and strongly compressed • A single unnotched dorsal fin decreasing in sizewith age; spinous part not well differentiated, with 3 to 9 spines increasing in size posteriorly and nearly concealed by the scaly covering of the fin • Short rounded, pectoral fins, pelvic fin, greatlyproduced. 	• Live in shallow protected coastal waters to deep, somewhat silty habitats. Often with deep shipwrecks.

No	Family Name	Key characters	Other features
3.	**Acanthuridae** (Surgeonfishes, Tangs and Unicorn fishes – 6genus and 84species). 1. **Blackspot surgeonfish-** *Acanthurus bariene* 2. **Powderblue surgeonfish-** *A. leucosternon* 3. **Brown surgeon fish-** *A. nigrofuscus* 4. **Lined surgeonfish-** *A. lineatus.* 5. **Elongate surgeonfish-** *A. mata* 6. **Epaulette surgeonfish-** *A. nigricauda* 7. **Convict surgeonfish-** *A. triostegus* 8. **Doubleband surgeonfish-** *A. tennentii* 9. **Sailfin tang-** *Zebrasoma veliferum* 10. **Humpback unicornfish-** *Naso brachycentron* 11. **Spotted unicorn fish-** *N. brevirostris.* 12. **One knife fish-** *N. thynnoides.* 13. **Humpnose unicornfish-** *N. tuberosus.,*	• High bodied compressed fishes with a singlefolding lancet-like spine or one or two bonyplates usually bearing sharp keels on either sideof caudal peduncle. • Single dorsal fin. • Caudal fin truncate to lunate. 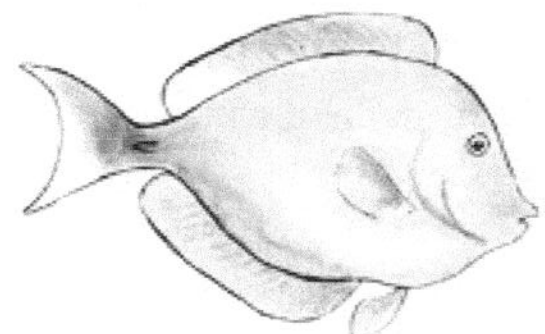	Acanthuridae are the family of surgeonfishes, tangs, and unicornfishes. The family includes about 86 extant species of marine fish living in tropical seas, usually around coral reefs. Many of the species are brightly colored and popular in aquaria. Distribution: Circumtropical, especially around coral reefs; five species in the Atlantic, the remaining in the Pacific and Indian oceans

No	Family Name	Key characters	Other features
4.	**Scatophagidae** (Scats – 2 Genus and 4sp). **1. Spotted scat-** *Scatophagus argus.*	• Highly compressed body, quadrangular shapedfishes • Head profile rising steeply to nape, mouth notprotrusible • Dorsal fin with 11 or 12 strong spines and 16 to 18soft rays, the first spine procumbent; a deep notchbetween spinous and soft portions of fin • Anal fin with 4 strong spines. 	• **Distribution:** Indo-Pacific. • Feed on algae and feces. Commonly keptin freshwater aquaria.
5.	**Siganidae** (Spine foots, Rabbit fishes – 1 genus and 15 sp). • **Streaked rabbit fish/ java rabbit fish-** *Siganus javus.* • **Gold lined spinefoot-** *S. lineatus.* • **White spotted spinefoot-** *S. canaliculatus.*	**Body laterally compressed** • Mouth small with a single row of fine, close-setteeth in each jaw • Dorsal fin with 13 strong spines and 10 soft rays, preceded by a forward projecting spine • Pelvic fins with 2 strong spines separated by 3 softrays, character unique to the family • Anal fin with 7 strong spines and 9 soft rays • The spines are venomous. 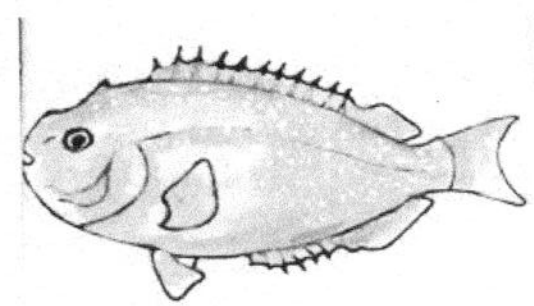	• Indo-Pacific and eastern Mediterranean. • Poisonous spines. • About 40 cm maximum length. • Some species in schools; others among corals. • All species are diurnal herbivores that feed on benthic algae. • Pelagic spawners. • Important foodfishes. • Some of the colorful species are popular in the aquarium trade.

No	Family Name	Key characters	Other features
6.	Zanclidae (Moorish idol-1 genera and 1sp). 1. Moorish idol - Zanclus cornutus (Linnaeus, 1758)	Body deep and compressed, to oval or discoid body Tubular snout with a small mouth containing numerous elongate bristle like teeth Dorsal spines elongated into a whip like filament Small horn like projection above eye, more prominent in adultsYellow in colour with two broad blackish brown cross bands, anterior band broader crosses the eye Caudal dark with yellow margin and narrow blue submarginal line. 	Indo-pan-Pacific. Monotypic. Common inhabitant of coral reefs.. Pelagic spawner; larvae drift for a long time before settlement, resulting in a wide distribution. Feed on mainly on sponges, also benthic invertebrates. Difficult to maintain (feed) in the aquarium.

Chapter-19

Order: Perciformes -VII

Marine Water Fishes

Order - Perciformes (Majority of species found in Marine water but few species Brackish water and Freshwater)

Kingdom: Animalia

Phylum: Chordata

Class: Actinopterygii (ray-finned fishes)

Order: Perciformes

Suborder: Scombroidei and Stromatodei

Family: Sphyraenidae, Gempylidae, Trichuridae, Xiphidae, Istiophoridae, Stromateidae

No	Family Name	Key characters	Other features
1.	**Sphyraenidae** (Barracudas – 1genra and 29sp)	• Body elongate and slightly compressed; long and pointed snout • Lower jaw projecting beyond upper jaw; strong canine teeth • Two short and widely separated dorsal fin, first with 5 fairly strong spines • Short pectoral fin • Forked caudal fin 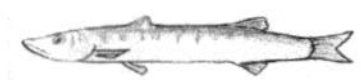	• Tropical and subtropical. Distribution: Atlantic, Indian, and Pacific Oceans. • Voracious predators of other fishes. • Attacks on humans have been reported. • Pelagic spawning in schools. • Food and game fishes, but large specimens may be ciguatoxic.
2.	**Gempylidae** (Snake mackerels, Barracudas – 16genus and 26 sp)	• Lower jaw projecting beyond tip of upper jaw • Detached finlets behind dorsal and anal fins • Pelvic fins usually small, often reduced to a single spine with only a few or no soft rays • No keels on caudal peduncle (except in Lepidocybium) • Lateral line single or double, ending at caudal fin base. 	• Usually found in very deep waters. Distribution: tropical and subtropical seas.

No	Family Name	Key characters	Other features
3.	**Trichiuridae** (Cutlassfishes / Hairtailfishes / Frostfishes / Scabbardfishes – 10 genus and 46sp)	• Body extremely elongate, compressed and ribbon like • Maxilla concealed by a preorbital bone • Dorsal fin low and long, beginning shortly behind eye • Anal fin low or reduced to short spines • Pectoral fins short and low on body • Pelvic fins reduced to a scale like spine (Benthodesmus sp.) or completely absent (in Trichiurus and Lepturacanthus) • Caudal fin small, forked or absent • Scales absent 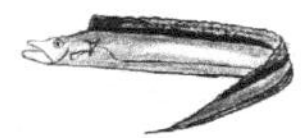	• Distribution: Atlantic, Indian, and Pacific
4.	**Xiphidae** (Swordfishes – 1genus and 1 sp) **Sword fish** - *Xiphias gladius* Linnaeus, 1758	• Body elongate and cylindrical • Upper jaw prolonged into a long bill (but both jaws prolonged into long bills in immature individuals) • No gill rakers on gill arches • Two separate dorsal fin in adults (continuous in immature specimens) the first much larger than the second • Pectoral fins falcate • Pelvic fins absent • A large keel on each side of caudal peduncle • Lateral line indistinct, but recognizable in specimens of 1 m length. 	Distribution: tropical and subtropical waters. High commercial value fish.

No	Family Name	Key characters	Other features
5.	**Istiophoridae** (Bill fishes, Spearfishes, Marlins and Sailfishes – 5 genus and 11 sp)	• Upper jaw prolonged into a long spear which is round in cross section • No rakers on gill arches • Two dorsal fins and two anal fins • 1 st dorsal fin and anal fin long, can both be folded back into grooves. • Pectoral fins falcate • Caudal fin large, strong and forked with a pair of keels on either side at base • Pelvic fin very long with 1 or 2 rays and a spine. 	• **Distribution:** most tropical and subtropical waters. • Bill used for stunning prey fish. • Very popular game fish.
6.	**Stromateidae** (Butterfishes, Silverpomfrets – 3 genus and 17 sp)	• Body very deep and compressed • Snout short and blunt; mouth small and immobile Maxilla • Single dorsal fin and anal fin; long-based and slightly to deeply falcate, preceeded by none or 5 to 10 flat, blade-like spines. • Pectoral fins long and wing-like • No pelvic fins. • Caudal fin forked; in some species with very extended lobes • Gill membranes broadly united to the isthmus. 	• **Coastal. Distribution:** North and south America, western Africa, and Indo-Pacific.

Family: Scombridae

Systematic classification

Kingdom: Animalia

Phylum: Chordata

Class: Actinopterygii

Order: Perciformes

Suborder: Scombrodei

Family: Scombridae

Identification of Family: Scombridae (Mackerals, Tunas and Bonitos)

- Body elongate and fusiform, moderately compressed in some genera.
- Snout pointed, premaxilla beaklike, free from nasal bones which are separated by the ethmoid bone; mouth large; teeth in jaws strong, moderate, or weak; no true canines; palate and tongue may bear teeth. The 2 dorsal fins separate and depressible into grooves with 5-12 finlets behind second dorsal and anal fins; first dorsal fin with 9-27 rays, origin well behind the head.
- Pectoral fins high on body. Pelvic fins moderate or small with 6 fin rays, placed below the pectoral fins.
- Caudal fin deeply forked with supporting caudal rays completely covering hypural plate. At least 2 small keels on each side of caudal fin base, a larger keel in between on caudal peduncle in more advanced species.
- Lateral line simple.

Layout

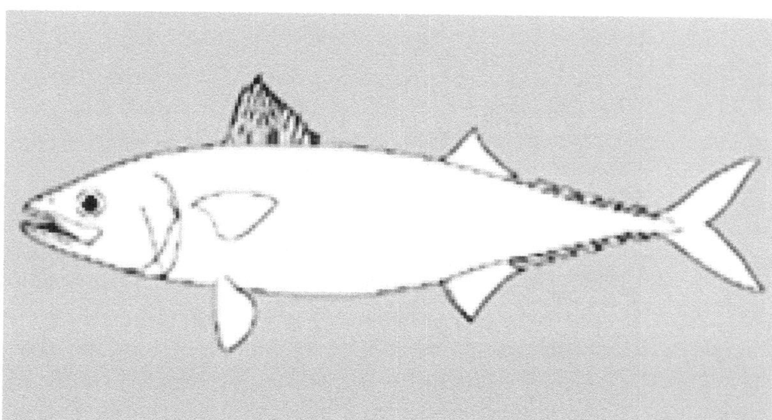

Chapter-20

Order: Perciformes

- ☆ The body is flattened
- ☆ Eyes are present in one side of the body.
- ☆ Body is not bilaterally symmetrical.
- ☆ Dorsal and anal fins with strong spines.
- ☆ Pelvic fin small.
- ☆ Dorsal fin overlaps the neurocranium.

Kingdom: Animalia

Phylum: Chordata

Class: Actinopterygii

Order: Pleuronectiformes

Family: Pesttodidae, Citharidae, Pleuronectidae, Soleidae, Bothidae and Cynoglossidae.

Features	Bothidae	Citharidae	Cynoglossidae	Pleuronectidae	Pesttodidae	Soleidae
Eyes position	Left side of the body	Left or right side of the body	Left side of the body	Right side of the body	Left or right side of the body	Right side of the body
Distribution	Tropical and temperate Atlantic, Indian, and Pacific oceans. Both eyes on left side in most species	Mediterranean, Indian Ocean and Japan to Australia.	Tropical and subtropical areas.	Chiefly marine;oc casionally brackish; rare in freshwater. Distribution: Arctic, Atlantic, Indian and Pacific Oceans.	West Africa and Indo-West Pacific	Primarily Europe to Australia and Japan.
Key features	• Preopercle margin free and distinct. • No spines in fins. • Dorsal fin origin above or anterior to upper eye. • Dorsal and anal fins separate from caudal fin. membranes united. • Anus on blind side. • Branchiostegal	• A single spine and five soft rays in pelvic fins; pelvics equally • developed and short based. • Dorsal fin origin at posterior nostril of blind side; pectoral • fins well • developed. • Branchiostegal membranes not united. • Lateral line present on both sides. • Anus on eyed side.	• Eyes very • small with little • interorbital space. • Preopercle margin • covered with skin. • Asymmetrical mouth. • Caudal fin • pointed and confluent with dorsal and caudal fins. • Right pelvic fin usually undeveloped. • No pectoral fins	• Preoperculum with a free margin. Fins without spines. • Dorsal fin extends on head. • Swimbladder absent in adults.	• Largemouth; vomer, palatine and tongue with minute teeth. • Distinct preopercular margin, not hidden beneath skin. Dorsal fin origin well behind eyes, the anterior rays spinous. • Almost symmetrical • pelvic fins with a weak spine and five soft rays. Cycloid scale; lateral line developed on both sides of body.	• Margin of preoperculum completely concealed (covered by skin and scales). • Dorsal fin origin above or before eyes; pectoral fins present or absent; no fin spines. • Dorsal and anal fins free from caudal fin or united with caudal fin. • Pelvics free from anal fin.

Features	Bothidae	Citharidae	Cynoglossidae	Pleuronectidae	Pesttodidae	Soleidae
Other features	• Predators of benthic invertebrates and fish-es. • Pelagic spawners. Egg yolk with a single oil globule. • Lefteye flounders are found in all tropical and temperate seas, and several species are of commercial importance		• Most species occur in shallow water or estuaries; • Symphurinae are in deep water of 300-1,900 m; many species are commercially important. • Five species are known primarily from rivers.	• The pigmented side is capable of remarkable color changes in order to match the bottom. • Predators of benthic invertebrates and fishes. • Pelagic spawners. • Eggs without oil globule in yolk. • Benthic, from a few to more than 1,000 m.	• Psettodids are considered the most primitive of the flatfishes, • as expressed, e.g., by few vertebrae and their habit of often swimming in an upright position. • Psettodes erumei reaches up to 64 cm)	• Soles are benthic, neritic fishes feeding on benthic invertebrates and fishes. • Pelagic spawners. • Oviparous • Eggs are pelagic and non-adhesive
Common name and scientific name.	1. **Indo-pacific oval flounder** – *Bothus myriaster.* 2. **Leopard flounder-***Bothus pantherinus.* 3. **Large-crested lefteye flounder-** *Arnoglossus tapieinosoma.* 4. **Broadbrow flounder-** *Crossorhombus valderostratus.* 5. **Threespot flounder-** *Grammatobothus polyopthalmus.*	1. **Large scaleflounder-** *Parachitharus macrolepis.*	1. **Large scaled tongue sole-** *Cynoglossus arel.* 2. **Round head toungesole-** C. dispar. 3. **Short headed tongue sole-** *C. kopsii.* 4. **Long tongue sole-** *C. lingua.* 5. **Speckled tongue sole-** *C. puncticeps.*	1. **Peppered flounder-** *Paralichthodes algoensis.*	1. **Indian halibut-** *Psettodes erumei.*	1. **Orintal sole-** *Brachirus orientalis.* 2. **Eyed sole-** *Heteromycterisoculus.* 3. **Kaup's sole-** *Synaptura albomaculata.* 4. **Commerson's sole-** *S. commersonnii.* 5. **Fringefin zebra sole-** *Zebrias quagga.* 6. **Indian zebra sole-** *Z. synapturoides.*

Features	Bothidae	Citharidae	Cynoglossidae	Pleuronectidae	Pesttodidae	Soleidae
	6. **Many-scaled flounder-** *Parabothus polylepis.*					
6. Layout	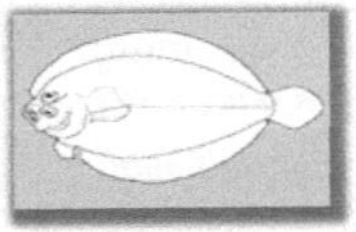	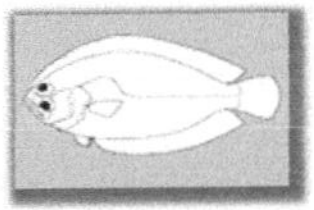	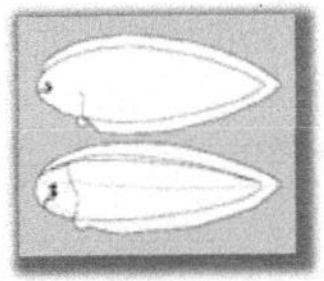	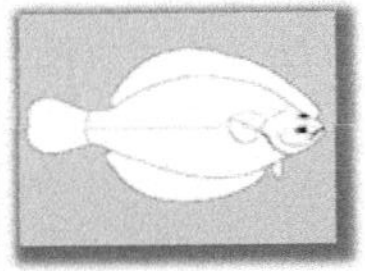		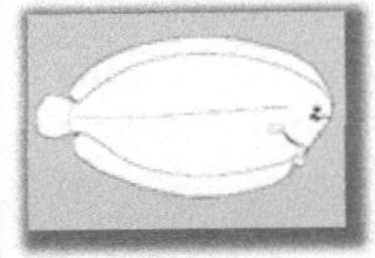

Chapter-21

Order: Siluriformes (Marine Water)

Morphological Characteristics

Mesopterygoid Reduction: Reduced mesopterygoid bone.

Opercle and Interopercle: Relatively small in size.

Adipose Fin: Usually present.

Spine-like Rays: Found at the front of dorsal and pectoral fins.

Fins and Skeletal Features

Dorsal Fin: Typically, with two spines; the first spine is very short and functions as a locking mechanism for the second spine.

Caudal Fin: Usually with 18 or fewer rays, often around 17. Variability in caudal skeleton, ranging from six separate hypural plates to complete fusion of caudal elements.

Body and External Characteristics

Body Covering: Naked or covered with bony plates.

Barbells: Up to 4 pairs present on the head; variation in presence/absence of nasal and chin barbels.

Maxilla: Typically, toothless and rudimentary, with exceptions like Diplomystidae and the extinct Hypsidoridae.

Eye Size: Generally small.

Adaptations and Behavior

Air Breathing Organs: Present in certain families like Claridae and Heteropreustidae.

Size Variation: Many species below 12 cm in length, with exceptions like Silurus glanis, reaching up to 3m.

Exceptionally Large Species: Pangasiid and Pimelodid known to reach large sizes.

Taxonomy and Distribution

Siluriformes Statistics: Families: 34; **Genera:** 412; **Species:** 2,405. Approximately 1,440 species are currently known.

Habitat Distribution: Ariidae and Plotosidae largely comprise marine species, with representatives found in brackish, coastal waters, and occasionally in freshwater. Other families are predominantly freshwater, with some species capable of inhabiting brackish water.

Commercial Importance: Ariidae and Plotosidae are commercially significant in marine environments

Kingdom: Animalia

Phylum: Chordata

Class: Actinopterygii

Order: Siluriformes

Family: Plotosidae and Ariidae.

Family	Key characters	Other features
1. Ariidae (Sea catfishes- 31 genra and 157sp). 1. **Spotted catfish-** *Arius maculatus.* 2. **Shovelnose sea catfish-** *A. subrostratus.* 3. **Blackfin sea catfish-** *A. jella.* 4. **Giant catfish-** *Netuma thalassinus.*	• Snout and head rounded to depressed • 1-3 pairs of barbels present • Head covered with a bony shield • A short adipose dorsal fin present • First dorsal fin short with a short spine or buckler. • Two pairs of adjacent nostrils on each side of snout. 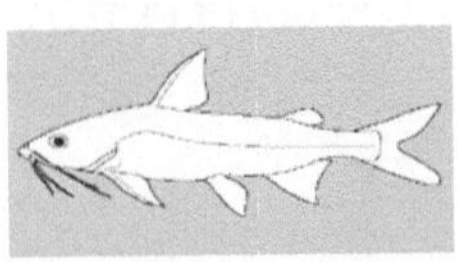	• Chiefly marine; occasionally freshwater. • **Distribution:** tropical and subtropical waters. • Many species enter freshwater and some are restricted to freshwater. Normally the male carries the relatively large eggs in its mouth until hatching.

Family	Key characters	Other features
2. Plotosidae *(Eeltail catfish- 10 genus and 42sp).* **1. Gray eel catfish-** *Plotosus canius.* **2. Striped eel catfish-** *Plotosus lineatus.*	• Elongate body, compressed, tapering to a point posteriorly • 4 pairs of barbels present with 1 pair nasal, 1 pair maxillary and 2 pairs mental; • First dorsal fin short-based with a serrated spine and 4-6 soft rays; second dorsal fin (or dorsal procurrent caudal fin), caudal fin and anal fin confluent • Absence of adipose fin • Pectoral fins with 1 serrated spine and 9 to 16 soft rays • A dendritic organ consisting of many vascularised epithelial folds present directly posterior to anus. • Caudal fin rounded or pointed. 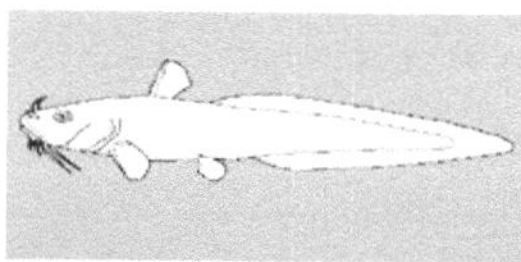	• **Distribution**: Indo-West Pacific (Japan to Australia and Fiji).

Gonorynchiformes, Berciformes, Aulopiforms, Elopiformes, Beloniformes and Mugiliformes

Marine water fishes - Order -Gonorynchiformes

Morphological Characteristics

- Orbitosphenoid Absent
- Teeth Absent on Fifth Ceratobranchial

Skeletal Features

- First Three Vertebrae Specialized
- Associated with One or More Cephalic Ribs
- Suprabranchial Organ Present
- Consisting of Lateral Pouches in the Posterior Part of the Branchial Chamber behind the Fourth Epibranchials

Oral and Dental Structure

- ❑ Small Mouth
- ❑ Jaws Toothless

Kingdom: Animalia

Phylum: Chordata

Class: Actinopterygii (ray-finned fishes)

Order: Gonorynchiformes

Family: Chanidae, Gonorynchidae, Kneriidae and Phractolaemidae

Family Name	Key characters	Other features
1. Chanidae *Milkfish- Chanos chanos* (Fabricius, 1775)	**Key Characteristics:** • **Body Shape:** Fusiform (elongated and tapering at both ends) and compressed. • **Mouth Position:** Terminal (at the end of the head). • **Jaw Structure:** Upper jaw is non protrusible. • **Scales:** Cycloid (smooth-edged scales). **Fins:** • Pelvic fin with 11 or 12 rays. • Dorsal fin with 13-17 rays. • Anal fin with 9-11 rays. • Four branchiostegal rays. • **Swim Bladder:** Present. 	• **Habitat:** Chiefly marine and brackish; sometimes found in freshwater. • **Distribution:** Indian and tropical Pacific Oceans. • **Size:** Can reach a maximum length of 1.7 meters, but typically about 1 meter. • **Diet:** Feeds on benthic algae and invertebrates. • **Economic Importance:** Valued as a food fish in Southeast Asia and extensively cultured in ponds. • **Fishing Method:** Young chanids are caught inshore to supply ponds for cultivation.

Marine water fishes - Order –Berciformes.

- ☆ **Orbitosphenoid Present:** This refers to the presence of a bone in the skull, specifically in the area near the eye sockets.
- ☆ **Two Supramaxillae Present:** Denotes the existence of two bones located above the maxillary bones, contributing to the structure of the skull.
- ☆ **Suboccular Shelf Present:** Indicates the presence of a shelf-like structure beneath the eye sockets, contributing to the overall framework of the skull.

Kingdom: Animalia

Phylum: Chordata

Class: Actinopterygii (ray-finned fishes)

Order: Berciformes

Family: Holocentridae

Family Name	Key characters	Other features
Holocentridae (Squirrel fishes, Soldier fishes – 8 genus and 90 species)	• **Pelvic Fin Spine:** Typically, a single spine exists with 5-8 soft rays (most commonly 7) in the pelvic fin. • **Dorsal Fin:** Long dorsal fin comprising 10-13 spines and a notched soft-rayed section with 11-17 rays. • **Anal Fin:** Anal fin with 4 spines, the third the stoutest and often the longest with Consists of four spines and 7-16 soft rays. • **Caudal Fin:** Forked caudal fin with principal caudal rays numbering 18 or 19. • **Scales:** Exhibits large and rough ctenoid scales. • **Eyes:** Large eye structure. • **Operculum:** Possesses a spiny-edged operculum. • **Vertebrae:** Typically has 26 or 27 vertebrae. • **Coloration:** Usually reddish in color. 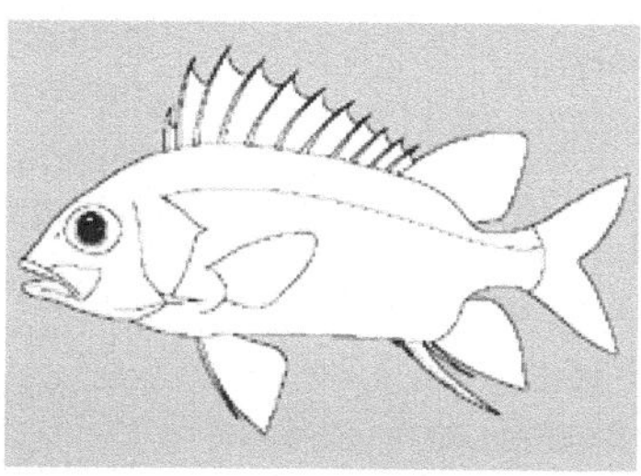	• **Distribution:** Found in tropical Atlantic, Indian, and Pacific Oceans. **Habitat and Behavior:** • **Nocturnal behavior;** often cryptic during the day, hiding in crevices or beneath reef ledges along with other species like cardinal fishes, bigeyes, and sweepers. • Primarily inhabits shallow waters, ranging from the shoreline to depths of 100 meters, rarely found beyond 200 meters. • **Feeding habits vary between species:** Soldiesfish primarily feed on large zooplankton, while squirrelfish feed on benthic invertebrates and small fishes. • Adults are usually demersal, staying near shelter on the bottom. Larvae are planktonic for several weeks and settle at a size of approximately 3 cm. • **Reproduction:** Myripristis species have been observed spawning in open water, typically a few days after a full moon. • **Size:** Maximum length recorded is about 61 cm, observed in Sargocentron spinifer. • **Aquarium Suitability:** Considered hardy in aquariums. • **Economic Importance**: Significantly valued as subsistence food fishes in various regions.

Marine Water Fishes Order – Aulopiforms.

Kingdom: Animalia

Phylum: Chordata

Class: Actinopterygii (ray-finned fishes)

Order: Aulopiforms.

Family: Harpadontidae and Synodontidae.

Family Name	Key characters	Other features
Harpadontidae (Bombay-duck). **Bombayduck -** *Harpadon nehereus* (Hamilton, 1822)	• **Dorsal Fin with Adipose Fin:** A prominent dorsal fin followed by a noticeable adipose fin. • **Pelvic Fins:** Remarkably long pelvic fins. • **Lateral Line Extension:** The lateral line extends into a pointed median lobe of the caudal fin. 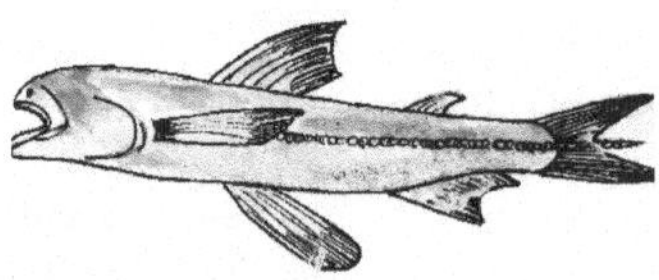	• Inhabit deep water offshore on sandy mud bottom for most of the year, but also gathers in large shoals in deltas of rivers to feed during monsoons. • Primarily caught along Maharashtra with the bag-net, better known as 'dol' net. Operation of this gear is timed to a strong tidal current. • Excellent food fish. Marketed fresh and dried or salted; consumed panfried
Synodontidae (Lizardfishes – 4 genus and 81 species).	• Species of Synodontidae (Lizard fishes) have the lateral line confined to the body and not extending as a median lobe of caudal fin **Physical Description:** • Small, cylindrical body structure. • Fins lack spines. • Large mouths equipped with slender, sharp teeth, including those on the tongue. • Elongate and cylindrical body with adipose fin. • Pointed teeth visible even when mouth is closed. 	• **Habitat:** Primarily marine, occasionally found in brackish waters. • **Distribution:** Found in the Atlantic, Indian, and Pacific Oceans. • **Feeding Behavior:** Voracious predators, primarily targeting small fishes as their prey. • **Supramaxilla:** If present, the supramaxilla is small. • **Branchiostegal Rays:** Typically possess 8-26 branchiostegal rays. • **Vertebrae:** The count of vertebrae ranges between 39-67. • **Reproductive System:** Generally non-hermaphroditic, except in the monoecious Bathysaurus.

Marine Water Fishes Order – Elopiformes.

Kingdom: Animalia

Phylum: Chordata

Class: Actinopterygii (ray-finned fishes)

Order: Elopiforms.

Family: Elopidae and Megalopidae

Order	Family	Characteristics	Example Species	Other features
ELOPIFORMES • well-developed gular plates, or extra bones in the throat between the lower jaws: 1. **Elopidae** (Ladyfishes, Tenpounders) 2. **Megalopidae** (Tarpons)	**Elopidae** (Tenpounders - 1genus and 7sp) 	• Well-developed gular plates between the lower jaws • Single dorsal fin with origin slightly behind the midpoint of the body	**Tenpounders** *-Elops machnata*	• Chiefly marine; enter estuaries and freshwater, especially as juveniles. Distribution: tropical and subtropical waters. • Larval stage translucid. Gamefish.
	Megalopidae (Tarpons-1genus and 2sp) 	• **Large eye** • Gular plate located between arms of lower jaw • Single dorsal fin present at the midpoint of the body; last dorsal fin ray filamentous	• **Indo pacific Tarpoons** - *Megalops cyprinoides*	• Chiefly marine; entering fresh water. Distribution: tropical and subtropical waters. • Larval stage leptocephalic translucid. Large silvery fishes. Can tolerate oxygen-poor water by inhaling air into lung-like swim bladder. Popular sportfishes.

Marine Water Fishes Order – Beloniformes.

Kingdom: Animalia

Phylum: Chordata

Class: Actinopterygii (ray-finned fishes)

Order: Beloniforms.

Family: Belonidae, hemiramphidae and Exocoetidae

Order	Familys	Example Species	Other Features
1. Beloniformes	**Belonidae** (Needle fishes – 10 genera and 47 sp) • Elongate fishes with both upper and lower jaws extended into a long beak • Dorsal and anal fins posterior in position • Pelvic fins located in abdominal position	**1. Banded needle fish -** *Strongylura leiura,* **2. Hound fish** - *Tylosurus crocodilus crocodiles.*	• **Distribution:** Marine and freshwater. Tropical and temperate waters. • **Interest to Fisheries:** Caught by casting or trolling surface or near-surface lures and in floating gill nets. Flesh excellent in flavor although some people have misgivings about eating it due to the presence of many small bones and the green color of the bones • **Habitat and Biology:** Most species are marine, but some occur in freshwaters. Carnivorous, feeding largely on small fishes which they catch sideways in their beaks. Needlefishes tend to leap and skitter at the surface and some people have been injured when accidentally struck by them, particularly at night when the fishes are attracted by lights. Eggs have entangling tendrils which act as holdfasts to each other and to objects in the water.

Order	Familys	Example Species	Other Features
2.	• Hemiramphidae (Halfbeaks-8genus and 61sp) • Elongate fishes with a prolonged lower jaw (except Oxyporhampus) and a short triangular upper jaw • Dorsal and anal fins posterior in position; pelvic fins in abdominal position • Lateral line running down from pectoral fin origin and then backward along the ventral margin of the body 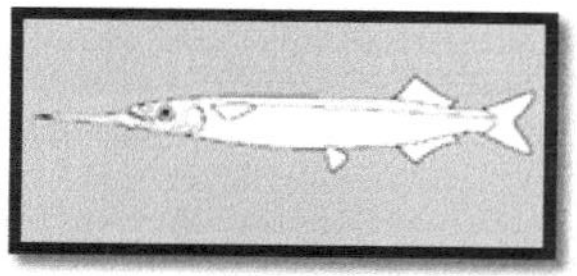	1. **Jumping Halfbeaks** *-Hemiramphus archipelagicus*	**1. General Information** • Halfbeaks are known from the Atlantic, Indian and Pacific Oceans. • Eggs attached to algae in shallow waters. • Viviparity in Dermogenys, Nomorphamphus, and Hemirhamphodon. • Surface-dwelling omnivores feeding on algae, zooplankton, and fishes. **2. Ecological and Behavioral Traits:** • Surface-dwelling and protectively colored for their mode of life. • Primarily marine, with some species inhabiting freshwaters. • Omnivorous, feeding on sea grasses, crustaceans, and small fishes. • Known for leaping and skittering at the surface; some species can glide like flying fish. **3. Economic Importance:** • Excellent flesh; larger species utilized as food globally. • Important as baitfish in the Gulf of Mexico and Caribbean Sea.

Order	Familys	Example Species	Other Features
3.	**Exocoetidae** (Flying fish - 7 genera and 71sp) • Pectoral fins high on sides, strikingly long, always extending beyond dorsal fin origin • Pelvic fins abdominal in position and greatly enlarged in many • Caudal fin deeply forked with lower lobe longer than the upper. 	**Large scale flying fish -** *Cypselurus oligolepis*	**4. Genera and Species:** • 12 genera and over 100 species in Hemiramphidae. • Zenarchopterinae recognized as a separate family, Zenarchopteridae, with about 50 sexually dimorphic species. • Notable genera include Hemirhamphus, Rhynchorhamphus, and Hyporhamphus **5.Geographic Distribution:** • Worldwide distribution with marine, estuarine, and freshwater species. • Western Atlantic halfbeak fauna includes nine marine species in five genera.

Marine Water Fishes Order – Mugiliformes.

Kingdom: Animalia

Phylum: Chordata

Class: Actinopterygii (ray-finned fishes)

Order: Mugiliiforms.

Family: Mugilidae

Mugilidae (Mullet -26 genra and 28 sp) **(Grey Mullet –** *Mugil cephalus*)

Distribution: All tropical and temperate seas. Chiefly marine (coastal) and brackish water; some in freshwater (Liza abu only in fresh and estuaries).

- **Adipose Eyelid:** Presence of an adipose eyelid.
- **Head Structure:** Flattened head, contributing to distinctive morphology.
- **Dorsal Fins:** Usually equipped with two dorsal fins.
- **Body Coloration:** Silvery body, providing a characteristic appearance.
- **Lateral Line:** Not present.
- **Body Shape:** Elongated body structure.
- **Mouth Size:** Relatively small mouth.
- **Pectoral Fin Placement:** Pectoral fin positioned high on the body.
- Travel in schools and feed on fine algae, diatoms, and detritus of bottom sediments. Important food fishes.

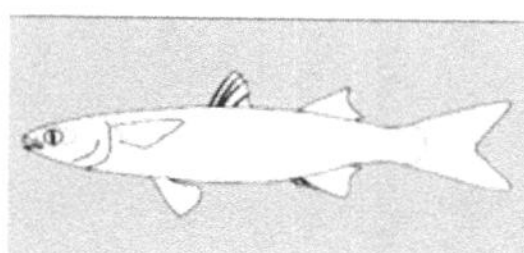

Tetradontiformes, Anguilliformes and Sygnathiformes

Marine water fishes

Order - Tetradontiformes (Majority of species found in Marine water but few species Brackish water and Freshwater)

- Pelvic fin absents or greatly reduced
- Usually joint into a bitting plate
- Small Gill Opening
- Skin thicker or Rough
- Sometime with spine scaly plates

Kingdom: Animalia

Phylum: Chordata

Class: Actinopterygii (ray-finned fishes)

Order: Tetradontiformes

Family: Tricanthidae, Balistidae, Monocanthidae, Ostracidae, Tetraodontidae and Diodontidae

Family Name	Key characters	Other features
Tricanthidae (Tripod fishes, Triplespines – 4 Genera and 7 Species) 1. **Long spined tripod fishes** - *Pseudotriacanthus strigilifer* 2. **Shortnose tripod** – *Triacanthus biaculeatus.*	**Body Morphology:** • Deep Compression • Shagreen-Like Skin Appearance **Dorsal Fin Features:** • Dorsal Fin commonly having 5 Spines, Rudimentary 6th Spine **Caudal Fin and Peduncle:** • Deeply Forked Caudal Fin • Tapered Caudal Peduncle **Pelvic Fin Representation:** • Large Spine as Pelvic Fin 	• **Distribution:** Indo-Pacific. Habitat: continental shelves; usually just below the sea surface to 60 m depth.

Family Name	Key characters	Other features
Balistidae *(Trigger fishes and Durgons -12 genera and 42 sp)* **1. Starry trigger fish** – *Abalistes stellatus* **2. Orange lined trigger fish** – *Balistapus undulatus* **3. Titan trigger fish** –*Balistades viridescens* **4. Sargassum trigger fish** – *Xanthichthys ringens* **5. Yellowmargin trigger fish** – *Pseudobalistes flavimarginatus* **6. Red-toothed trigger fish** – *Odonus niger* **7. Halfmoon trigger fish** – *Sufflamen chrysopterum* **8. Masked trigger fish** – *Sufflamen fraenatus* **9. Wedgetail trigger fish** – *Rhinecanthus rectangulus.*	**Body Structure:** • Deep and Moderately Compressed Body • Very Thick, Tough Skin **Scales and Tympanum:** • Tympanum Formed by Scales Above Pectoral Fin Base **Dorsal Fin Features:** • Two Dorsal Fins • First Dorsal Fin with 3 Spines • Second Dorsal Fin with Branched Fin-Rays **Spine Characteristics:** • Relationship Between First and Second Dorsal Fin Spines • Locking Mechanism of First Spine **Pelvic Fins and Scales:** • Rudimentary or Absent Pelvic Fins • Enlarged Scales Encasing Pelvic Region. 	• **Distribution:** Atlantic, Indian and Pacific Oceans. Habitat: shallow waters, mainly on coral reefs around the world, usually from just below the sea surface to 50 m depth. • Triggerfish normally swim by undulating their second dorsal and anal fins, but will use their tail for rapid bursts. • Most triggerfishes are solitary diurnal carnivores, feeding on a wide variety of invertebrates including hard-shelled mollusks and echinoderms; some also feed on algae or zooplankton. • They lay demersal eggs in a nest which is aggressively guarded by the female, less often by the male. • Popular and hardy in aquaria, but often aggressive

Family Name	Key characters	Other features
Monocanthidae (Filefishes / Leather jackets- 27 Genera and 108 sp). 1. **Japanese leather jacket** – *Paramonocanthus nipponensis.* 2. **Unicorn leather jacket** – *Aluterus monoceros* 3. **Scrawled filefish** – *Aluterus scriptus.* 4. **Bearded leather jacket** – *Anacanthus barbatus.* 5. **Rhino leather jacket** – *Pseudalutarius nasicornis.*	**Skin Texture and Scale Structure:** • Smooth to Rough Shagreen-Like Skin • Minute Scales with Fine Spicules **Dorsal Fin Spines:** • Two Dorsal Fin Spines • Large First Spine with Downward-Curving Barbs • Locking Mechanism of the First Spine with the Small Second Spine **Pelvic Fins and Scales:** • Rudimentary Pelvic Fins • Enlarged Scales Encasing the Pelvic Region • Variability in the Presence of Pelvic Fins with Age. 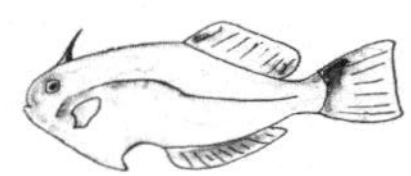	• Distribution: Atlantic, Indian and Pacific Oceans. • Most species feed on a wide variety of benthic invertebrates, but some specialize on corals or zooplankton. • Filefishes lay demersal eggs in a site prepared and guarded by the male or both parents. • Some of the subtropical species release eggs in open water.,
Ostracidae *(Box fishes/ Cow fishes – 7 genera and 25sp).* 1. **Longhorn cowfish** – *Lactoria cornuta* 2. **Yellow box fish** – *Ostracion cubicus* 3. **Humpback turret fish** – *Tetrosomus gibbosus*	• Body almost completely encased in a bony shell, or carapace formed of enlarged, thickened scale plates • Absence of pelvic fin. • Dorsal fin rays 9-13. Anal fin rays 9-13. • Non-protrusible upper jaw. 	**Distribution:** Atlantic, Indian and Pacific Oceans. • ome species of trunkfishes secrete ostracitoxin, poisonous to other fishes and, to some extent, even to trunkfishes. • Feed on small sessile invertebrates and on algae. They are territorial and haremic, spawning pelagic eggs at dusk. Not recommended for aquaria. Other common names in use are cowfishes and trunkfishes.,

Family Name	Key characters	Other features
Tetradontidae (Puffer fishes/Blow fishes/ Tobies – 28 genera and 205 sp) 1. **Blue spotted buffer** – *Arothron caeruleopunctatus* 2. **Spotted sharpnose/ Slender toby** – *Canthigaster solandri.* 3. **White spotted puffer** – *Arothron hispidus.* 4. **Immaculate puffer** – *A. immaculatus.* 5. **Black spotted puffer** – *A. nigropunctatus.* 6. **Giant Puffer fish/ Starry toad fish** – *A. stellatus.* 7. **Smooth blassop** – *Lagocephalus inermis.* 8. **Lunartail puffer** – *L. lunaris.* 9. **Lattice blassop** – *Takifugu oblongus* 10. **Short fin puffer** – *Torquigener brevipinnis.*	**Body Structure and Inflation:** • Heavy and Blunt Body • Rapid Inflation Mechanism through Water or Air Intake **Head and Jaws Adaptations:** • Large and Blunt Head • Jaws Modified to Form a Beak • Heavy, Powerful Teeth (2 Above and 2 Below) • Eye Placement: • Eyes Located High on Head **Pelvic Fins Absence:** • Absence of Pelvic Fins • Fin Placement and Spine Absence: • Dorsal and Anal Fins Located Far Posteriorly • No Dorsal Fin or Anal Fin Spines **Scale and Prickle Characteristics:** • Absence of Typical Scales • Numerous Small Spiny Prickles on Back and/ or Belly • Occasional Presence of Prickles on Sides 	• **Distribution:** tropical and subtropical areas of Atlantic, Indian and Pacific. Habitiat: Chiefly marine. Many going into and inhabiting brackish and freshwater. • Some puffers contain tetraodotoxin, especially in the viscera; in the gonads of some during spawning season. • **Food:** Some species seem to consume about everything, whereas others have preferences for certain invertebrates or algae. • **Life history:** Demersal eggs are laid in a nest and presumably defended. • **Management:** European Community legislation prohibits trading with puffer fish products

Family Name	Key characters	Other features
6. Diodontidae (Porcupine fishes, Spiny puffers, Baloon fishes - 7 genera and 18 sp). 1. **Bird beak burr fish** – *Cyclichthys orbicularis.* 2. **Longspine porcupine fish** – *Diodon holocanthus.* 3. **Spotfin porcupine fish** – *D. hystrix.*	**Body Morphology and Inflation:** • Wide Body Capable of Great Inflation • Presence of Spines Covering the Body skins. **Dental Structure:** • Teeth Fused to Form Beak-Like Crushing Structure • Division of Upper and Lower Jaws into Left and Right Halves **Pelvic Fins Absence:** • Lack of Pelvic Fins **Lateral Line and Scale Features:** • Inconspicuous Lateral Line • Absence of Normal Scales 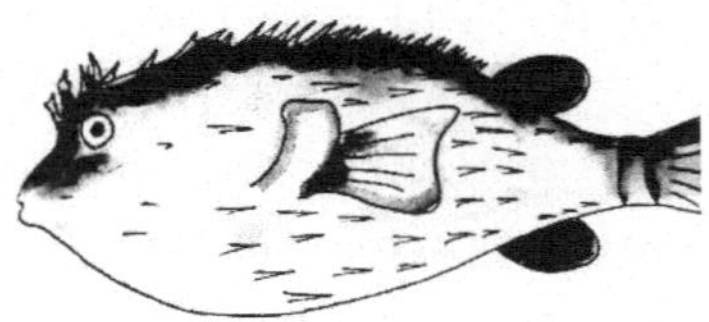	• **Distribution:** Atlantic, Indian, and Pacific Oceans. • Adults found inshore; the young pelagic. • Feed primarily on hard-shelled invertebrates crushed by the beak. Eggs are pelagic.

Marine Water Fishes

II. Order - Anguilliformes (Majority of species found in Marine water but few species Brackish water and Freshwater).

- Pelvic fins and supporting bones absent
- Pectoral fins absent in some fishes
- Skeleton lack bony connection to skull
- Anal fin and dorsal fin join with caudal fin
- Scales are usually absent
- Gill rakers absent
- Large swim bladder present
- Body slender and elongate
- Leafy, transparent like leptocephali larvae
- Myomeres are more than 100 in larvae

Kingdom: Animalia

Phylum: Chordata

Class: Actinopterygii (ray-finned fishes)

Order: Anguilliformes

Family: Anguillidae, Muraenidae, Ophichthidae, Muraenesocidae, Nemichthyidae, Congridae and Xenocongridae.

Family	Scales	Jaw Comparison	Dorsal Fin Position	Mouth Size and Position	Gill Opening Size	Caudal Fin and Tail Tip	Supraorbital Pore	Pectoral Fin
1. Congridae	No scales	Lower jaw equal to or shorter than upper	Above or before pectoral tips	Normal size, not extending beyond eye	Normal size	Present	Absent	Present
2. Muraenesocidae	No scales	Mouth very large, beyond eye	dorsal fin join with caudal fin	Very large, extending beyond eye	Large	Present	Absent	Present
3. Ophichthidae	No scales	No caudal fin; tail tip hard and burrowing	dorsal fin join with caudal fin	Normal	Large	Tail tip a hard burrowing point	Present	Variable
4. Muraenidae	No scales	Small gill opening, small hole	dorsal fin join with caudal fin	Small, not extending beyond eye	Small	Absent	Absent	Absent
5. Xenocongridae	Reduced lateral line system	Gill opening a small hole	dorsal fin join with caudal fin	Normal	Small	Present	Absent	Variable

Family Name	Key characters	Other features
1. Anguillidae (Freshwater eels – 2 genera and 20sp). 1. **Indonesian Shortfin eel** - *Anguilla bicolor* 2. **Indian Mottled eel** - *A. bengalensis.*	**Scale Arrangement and Pattern:** • Small Oval Scales Embedded in Skin • Basket-Weave Pattern of Scale Arrangement **Body Morphology:** • Elongate and Snake-Like Body **Mouth and Jaw Structure:** • Lower Jaw Longer than Upper • Projecting Mouth, Angle behind Near Margin of Eye **Fins and Appendages:** • Absence of Pelvic Fin • Well-Developed Pectoral Fins **Anatomy and Positioning:** • Anus in Anterior Half of the Body • Vertical Slit-Like Gill Opening • Complete Lateral Line on Body and Head • Vertebrae Count (100s – 119). **Life Cycle and Migration:** • Catadromous and Spawning Migration • Leptocephali Metamorphosis into Elvers. 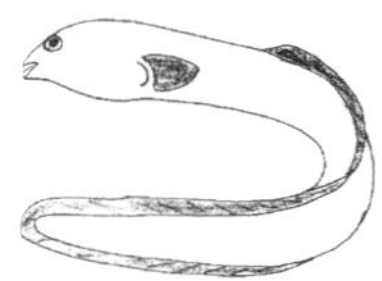	• Usually catadromous fishes in tropical and temperate waters, except eastern Pacific and south Atlantic.

Family Name	Key characters	Other features
2. Muraenidae *(Morray eel – (16 genera and 211 sp)* 1. **Laced moray** – *Gymnothorax favagineus* 2. **Yelllow edged moray** – *G. flavimarginatus.* 3. **Turkey moray/White mouth moray** – *G. meleagris.* 4. **Onespot moray**- *G. monostigma.* 5. **Slender giant moray** – *Sterophidon sathete* 6. **Peppered moray** – *G. pictus.* 7. **Indo pacific moray** – *G. randalli.* 8. **Undulated moray** – *G. undulatus.*	• Muraenidae, commonly known as moray eels. • Common Name: Moray Eel • Dorsal Profile: Steep above and behind the eye. • Gill Opening: Small, roundish lateral hole or slit. **Fins:** • Dorsal fin and anal fin are continuous around the tail. Pectoral and pelvic fins are absent. • Lateral Line Pores: No lateral line pores on the body, but a reduced complement on the head, including 1 or 2 above and before the gill opening (branchio lateral line pores). **Scales:** No scales. **Gill Arch:** Reduced gill arches. **Nostril Position:** Posterior nostril is high on the head. **Teeth:** Fang-like teeth are present. **Vertebrate Count:** Usually 110-200. **Genera:** Over 200 species reported under 15 genera. 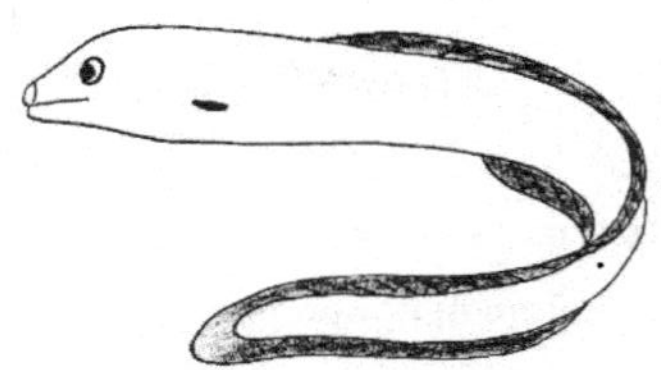	• **Distribution:** Worldwide in tropical and temperate seas. • Some morays cause ciguatera fish poisoning. A skin toxin was noted in an Indo-Pacific moray eel. • Adults benthic, generally in shallow water among rocks and coral heads; many species are more active at night and hide in holes and crevices during the day. • Vicious reputation is undeserved, although some species will bite if provoked. • Feed mainly on crustaceans, cephalopods and small fishes. • Larvae (leptocephali) epipelagic, widespread and abundant. Widely used as food, but a few large species may be ciguatoxic. • Generally hardy in captivity. Leptocephali showed curling behavior which make them resemble gelatinous zooplankton in observations made at Osprey Reef, Coral Sea.

Family Name	Key characters	Other features
3. Ophicthidae *(Worm eels and snake eels – 62 genera and 335 sp)* 1. ***Highfin snake eel*** *– Ophichthus altipennis.* 2. ***Evermann's snake eel*** *– O. lithinnus*	• The characteristics described align with the family Anguillidae, commonly known as freshwater eels. • **Body Shape:** Elongate, slender, and cylindrical. • **Gill Opening:** Small. • **Pectoral Fins:** Absent. • **Dorsal Fin:** Originating about a pectoral fin length behind the tips of the pectoral fins. • **Vertebrate Count:** 171-173 (approx.). 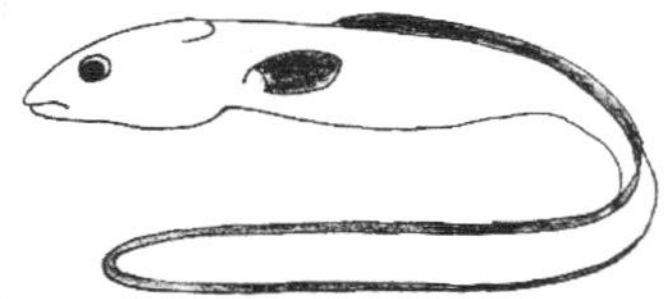	• **Distribution:** tropical to warm temperate waters. Habitat: shallow coastal waters to mid waters, reefs, rubble, sand and muddy substrates, usually < 400 m deep (although some Ophichthus were reported from 1,300 m depth. Some species entering rivers. • **Food/Diet:** most species spend their time burried in sand and hunt small fishes and crustaceans by sense of smell.
Muraenesocidae (pike congers-6Genus and 15sp). 1. **Common pike conger –** *Muraenesox bagio.* 2. **Daggertooth pike conger** *– M.cinereus.*	**Body Shape:** • Long and Cylindrical: The body is long to very long, more or less cylindrical in front. • Compressed Tail: The body is compressed along the tail. **Head and Mouth:** • Well-developed Mouth: The mouth is well developed and extends beyond the eye. • Terminal Mouth: The mouth is terminal, extending well beyond the eye.	**Distribution:** Atlantic, Indian and Pacific. **Tropical**

Family Name	Key characters	Other features
	Fins and Scales: • Dorsal Fin: The dorsal fin begins more or less above the gill opening, and the snout is very pointed. • No Pelvic Fin: There is no pelvic fin. • Pectoral Fins: Pectoral fins are present. • No Scales: The species lacks scales. **Teeth and Dentition:** • Well-developed Teeth: Teeth are well developed and fang-like. **Eyes and Snout** • Eyes large and covered with skin • Body very long with only tail compressed • Snout elongate • Vertebrae 120-216 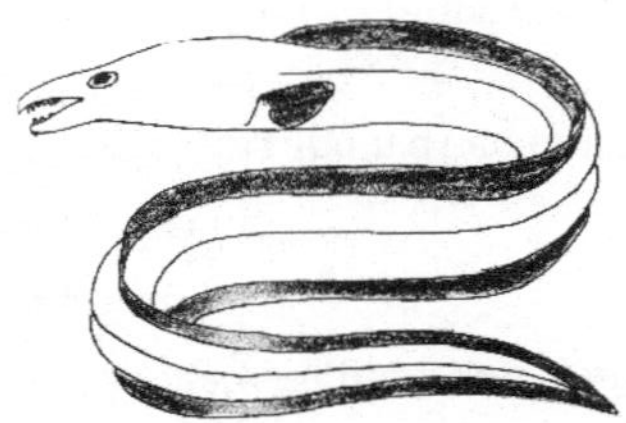	

Family Name	Key characters	Other features
Nemichthyidae (Snipe eels-3genus and 9sp): **1. Slender snipe eel**- *Nemichthys scolopaceus*	**Jaws and Mouth:** • **Extremely Long Jaws:** Jaws are extremely long, needle-like. • **Body Shape:** The body is long and slender. **Fins and Eyes:** • **Pectoral Fin:** Pectoral fin is present. • **Large Eyes:** Eyes are relatively large. **Anus Position:** • **Anus Location:** The anus is positioned far forward. 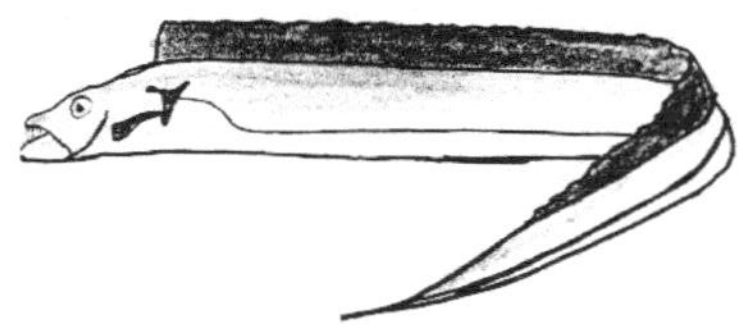	• **Distribution:** Atlantic, Indian, and Pacific. Bathypelagic and mesopelagic. Jaws extremely long; the lower jaw shorter than the upper making the jaws nonocclusible, with the exception of ripe males. • Sexually mature males undergo distinct changes, like the radical shortening of the jaws and tooth loss.

Family Name	Key characters	Other features
Congridae (Conger eels-29genus and 209 sp): 1. **Slivery eel** – *Ariosoma anago.* 2. **Cope conger** – *Conger wilsoni*	**Fin Structure:** • **Continuous Dorsal and Anal Fins:** Dorsal fin and anal fin are continuous around the tail. • **Dorsal Fin Origin:** Dorsal fin begins more or less above the gill opening and originates before pectoral fin tip. • **No Pelvic Fins:** There are no pelvic fins. **Body Characteristics:** • **Body Structure:** The body is robust without scales. • **Supratemporal Pore:** Supratemporal pore is present in front of dorsal fin. • **Lateral Line:** Lateral line is complete. **Mouth and Teeth:** • **Mouth Shape:** Mouth does not extend beyond the eye. • **Well-developed Teeth:** Teeth are well developed, but no canines in the jaw. • **Jaw Length:** Lower jaw is equal to or shorter than the upper jaw. **Vertebrae Count:** • **Vertebrae Count:** The number of vertebrae ranges from 205 to 225. 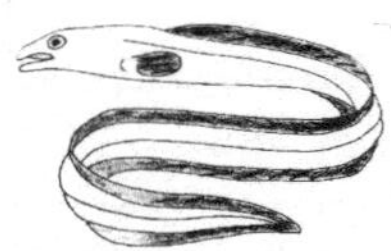 •	• **Distribution:** Atlantic, Indian and Pacific Oceans; mostly in deep or temperate waters and are valued as food fishes. • **Bahaviour:** Occur in colonies in sand where they live individually in burrows from which they protrude to feed on plankton; also feeds on small fishes and crustaceans, primarily at night. Mating and spawning apparently takes place without any lengthy migration. • Leptocephali (larvae) showed curling behavior which make them resemble gelatinous zooplankton, e.g. jellyfishes, in observations made at Osprey Reef, Coral Sea.

Marine Water Fishes

III. Order - Sygnathiformes

Kingdom: Animalia

Phylum: Chordata

Class: Actinopterygii (ray-finned fishes)

Order: Sygnathiiformes

Family: Sygnthidae, Solonocentridae, Centriseidae, Fistaluridae, Pegasidae and Aulostomidae

Characterstics	Sygnathidae	Solonocentridae	Centriseidae	Fistaluridae	Pegasidae	Aulostomidae
Body shape	Elongated, small	Relatively short	Thin razor like	Depressed and naked	Broad and depressed	Compressed, elongate, scaly
Mouth	Jointed	Large	Toothless	Tube like	Small mouth under a rostrum	Small fleshy barbels present
Head structure	Extended, separated	Large gill opening	Compressed	Naked and line of suites	Long rostrum present	Small isolateral spines
Dorsal fin	Present	2 dorsal fin(1 dorsal fin with 5 spines)	Sing with spines	Absent	Absent	Anal and second dorsal fin posterior
Anal fin	Short(2rays)	Present	Caudal and 2nd dorsal displaced ventrally	Forked with elongated	Very small	Absent
Pelvic fin	Absent	Present	Absent	Absent	Absent	Absent
Caudal fin	Absent	Present	Present	Present	Very small	Absent

Characterstics	Sygnathidae	Solonocentridae	Centriseidae	Fistaluridae	Pegasidae	Aulostomidae
Plates/scales	Numerous bony plates	-	Thin plates present	Tube shaped ossifications.	-	Scaly
Habitat	Shallow water	-	Swim vertical position	-	-	-
Broodpouch	Male will carry the brood pouch	Female will carry brood pouch	-	-	-	-
Species	1. Hippocampus sp 2. Hippichthys sp 3. Halicampus sp 4. Sygnathoides sp 5. Trachyrhamphus sp	Solenostomus	1. Centriseus sp	**Red Cornet fish** – *Fistularia petimba*	**Short dragon fish** – *Eurypegasus draconis.* **Longtail seamoth-** *Pegasus volitans.*	Aulostomus sp
Layout				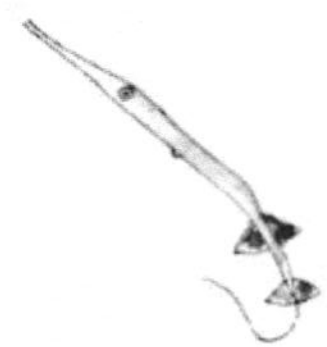		

Chapter-22

Order: Cypriniformes (Fresh Water)

Fresh Water Fishes

Order - Cypriniformes

Body Characteristics

- Body oblong and compressed.
- Small to large scales covering the body.
- Head lacks scales.

Bony Features

- Bony plates are not developed.
- Jaws, palatine, and pre-rygoid bones are toothless.

Mouth

- Mouth is usually protractile.
- Lack teeth in the jaws, palatine, and pre-rygoid bones.

Fins

- Single dorsal fin.
- Adipose dorsal fin is absent, except in some nemachelins and cobitins.

Barbels

- Barbels (whisker-like sensory organs) are present around the mouth, but they may be absent in some species.

Weberian Apparatus:

- Weberian apparatus, an anatomical structure associated with the swim bladder and hearing, is mostly modified in Cyprinidae.

Branchiostegal Rays:

- Branchiostegal rays are usually three, and there is no supra-branchial organ.

Distribution:

- Cyprinidae is a diverse family found in various aquatic habitats worldwide.

Economic Importance:

- Many species within Cyprinidae are economically important in aquaculture and as food sources.

Behavior:

- The family includes species with various ecological and behavioral adaptations, ranging from bottom-dwelling forms to those that occupy the open water.

Reproduction:

- Cyprinids often exhibit diverse reproductive strategies, including external fertilization and various spawning behaviors.

Examples:

- Common examples include common carp (*Cyprinus carpio*), goldfish (*Carassius auratus*), roach (*Rutilus rutilus*), and minnows like the fathead minnow (*Pimephales promelas*).

Kingdom: Animalia

Phylum: Chordata

Class: Actinopterygii (ray-finned fishes)

Order: Cypriniformes

Family: Cyprinidae (Carps, Minnows), Parapsilorhynchidae, Balitoridae, Cobitidae and Psilorhynchidae

Characteristics	Family – Cyprinidae	Subfamily – Leuciscinae	Subfamily - Danioninae	Subfamily – Cyprininae	Subfamily - Oreininae (Schizothoracinae)	Subfamily - Garrinae
Body	Scaled, generally compressed; abdomen rounded or cutting	Strongly keeled from breast to vent	Silvery, abdominal keel extends to pectoral fins level	Generally elongated or oval, compressed head without scales	Body spindle-shaped, ventral surface flat with scales	Inferior mouth, suctorial disc may be present
Eyes	Never covered with skin	Not applicable	Not applicable	Not applicable	Fairly prominent, lateral, visible from below ventral surface	Not applicable
Lips	Usually thin, not with papillae; may be absent from one jaw or closely infesting both	Absent; no maxillary foramen	Symphyseal knob on lower jaw may be present or absent	Mostly protractile and always toothless, jaws smooth or with a horny covering	Not applicable	Upper and lower lip may or may not be connected
Labial Fold	Continuous or interrupted, occasionally with cartilaginous covering	Not applicable	Not applicable	Not applicable	Not applicable	Upper lip continuous with skin, snout crenulated
Subocular Spines	Absent; Opercular bones well developed	Not applicable	Not applicable	Not applicable	Not applicable	Not applicable

Characteristics	Family – Cyprinidae	Subfamily – Leuciscinae	Subfamily - Danioninae	Subfamily – Cyprininae	Subfamily - Oreininae (Schizothoracinae)	Subfamily - Garrinae
Barbels	One or two pairs present or absent	Absent	Absent	One or two pairs present or absent	One or two pairs present or absent	May be absent
Dorsal Fin	Last undivided dorsal fin ray osseous or articulated; no adipose dorsal fin	Not applicable	Dorsal fin without osseous ray, usually with seven branched rays	Dorsal fin with or without spines, with 7 to 30 branched rays	Dorsal fin with or without spines	Not applicable
Anal Fin	Short, with 5 to 9 branched rays	Not applicable	Not applicable	Not applicable	Not applicable	Not applicable
Pharyngeal Teeth	One to three rows, never more than eight teeth in any one row	Not applicable	One to three rows	Always toothless	Two or three rows	Not applicable
Lateral Line	Complete or incomplete	Not applicable	Complete; curved downwards in lower half of caudal peduncle	Complete or incomplete, generally running in middle of the body	Not applicable	Not applicable

Characteristics	Family – Cyprinidae	Subfamily – Leuciscinae	Subfamily - Danioninae	Subfamily – Cyprininae	Subfamily - Oreininae (Schizothoracinae)	Subfamily - Garrinae
Air Bladder	Usually large, divided into anterior and posterior chambers, not surrounded by a bony capsule	Not applicable	Not applicable	Usually large, divided into anterior and posterior chambers	Not applicable	Not applicable
Species	Carp and Minnows (158genera and 1574sp)	**Genus:** Hypophthalmichthys **Species:** *i) H. molitrix* *ii) H. nobilis*	1. *Securicula gora* – only one species 2. Salmostoma spp – which consists of 12 species 3. Aspidoparia – consists of two species *A. jaya and A. morar* 4. Barilius spp – consists of 17 species 5. *Inlecypris auropurpureus* – only one species 6. *Chela* – consists of 4 species	Mentioned below	1. Schizothorax – Four species 2. Schizothoraichthys – 13 species 3. *Schizopygopsis stoliczkae* – only one species 4. *Diprychus maculates* – single species 5. *Phychobarbus corirostris* – single species	It consist of two Genus *1. Garra* sp – 38 species *2. Crosso cheilus* – 3 species

Characteristics	Family – Cyprinidae	Subfamily – Leuciscinae	Subfamily - Danioninae	Subfamily – Cyprininae	Subfamily - Oreininae (Schizothoracinae)	Subfamily - Garrinae
			a. C. Cachisus *b. C. fasciatus* *c. C. laubuca* *d. C. dadyburjori* **Sub genera** a. Chela b. Allochela c. Neochela 7. Esomus - which consist of 5 species 8. 8. Danio - which consist of 12 species 9. 9. Brachydanio - which consist of 7 species 10. 10. *Horadandia atukorali* - Only one species 11. 11. Microrasbora - This has 3 species 12. 12. *Danionella translucida* - only one species			

Characteristics	Family – Cyprinidae	Subfamily – Leuciscinae	Subfamily - Danioninae	Subfamily – Cyprininae	Subfamily - Oreininae (Schizothoracinae)	Subfamily - Garrinae
			13. Rasbora - consists of 7 species 14. *Rasboroides vaterifloris* - only one species 15. Bengala elanga - only one sp. 16. Amblypharyngodon - 5 species			

Cyprinidae Family		
Genus	**Number of Species**	**Common Name / Remarks**
Ctenopharyngodon Idella	1	Grass Carp
Cyprinus sp.		
C. carpio carpio	1	Common Carp
C. carpio communis	1	Scale Carp
C. carpio specularis	1	Mirror Carp
C. carpio nudus	1	Leather Carp
Carassius sp. (2sp)		
Carassius auratus	1	Goldfish
Carassius Carassius	1	Crucian carp
Tinca tinca	1	Tench
Thynnichthys sandkhol	1	Sandkhol carp
Capoeta capoeta	1	Seven Khramulya
Tor spp. – 9- Mahseer		
- *Tor khudree*	1	Deccan Mahseer
- *T. mosal*	1	Mahanadi Mahseer
- *T. mussullah*	1	Mussullah Mahseer
- *T. putitora*	1	Golden Mahseer
- *T. tor*	1	Tor barb
Naziritor spp. (2 sp)		
Neolissochilus sp	11	
Osteobrama sp	9	
Cyclochelichthys sp	1	
Rohree sp	1	
Mystacoleucus sp	2	
Poropuntius sp	6	
Chagunius sp	3	
Dreichthys sp	1	
Puntius (75 sp)		
Sawbwa sp	1	
Hampala sp	1	
Gononproktoptenus sp	10	
Semiplotus sp	4	King-fish
Scaphiodonichthys sp	1	
Osteochilichthys sp	4	

Cyprinion sp	2	
Kantaka brevidorsalis	1	
Labiobarbus sp	1	
Osetochilus sp	1	
Cirrhinus (5sp)		
- *Cirrhinus cirrhosis*	1	White Carp
- *C. fulungee*	1	Deccan White carp
- *C. mrigala*	1	Mrigal carp
- *C. reba*	1	Reba carp
- *C. macrops*	1	Hora White Carp
Catla catla	1	Catla
Labeo (28sp)		
- *Labeo rohita*	1	Rohu carp
- *L. konitus*	1	Pigmouth carp
- *L. dussimieri*	1	Common labeo/Malabar labeo
- *L. calbasu*	1	Orangefin labeo
- *L. bata*	1	Bata
Schismatorhynchos nukta	1	Nukta
Hgoralabiosa sp	3	

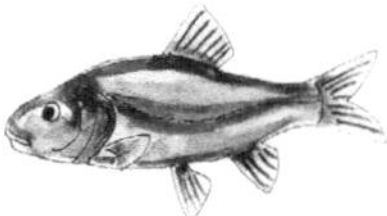

Cyprinidae

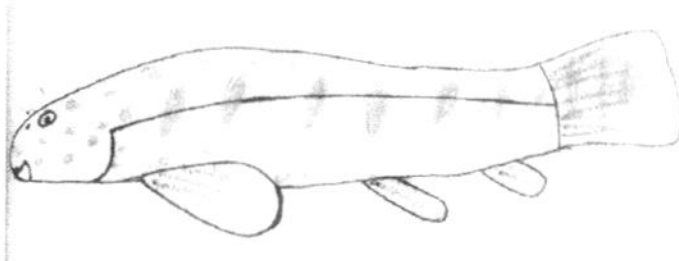

Balitoridae

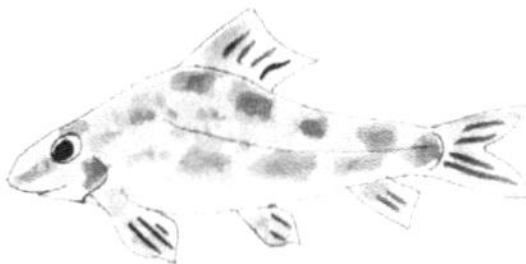

Psilorhynchidae

Parasylorhynchidae

Cobitidae

Family	Key Characters	Subfamily	Genus	Species
Parasylorhynchidae	1. Body spindle-shaped with flat ventral surface 2. Short, compressed head with pointed operculum 3. Very small, inferior mouth 4. Lower jaw with sharp edges and prominent lips 5. Single rostral pair of barbels 6. Fairly prominent lateral eyes 7. Anal fin with 5 branched rays 8. Dorsal fin with 8 branched rays		Parasylorhynchus	1. *P. discophorus* 2. *P. prareri* 3. *P. tentaculatus*
Psilorhynchidae	1. Flattened body with small scales 2. Small, inferior mouth with projecting snout 3. Narrow gill opening with united membranes 4. No pseudobranchiae 5. Slender pharyngeal bones with four teeth 6. Absence of barbels 7. Pectoral and pelvic rays vary 8. Complete lateral line, reduced air bladder		Psilorhynchus	1. *Psilorhynchus* spp (four species)
			Psilorhynchoides	2. *Psilorhynchoides* spp (three species)

Family	Key Characters	Subfamily	Genus	Species
Cobitidae	1. Body short, low, more or less cylindrical, posteriorly compressed but never depressed 2. Body with small cycloid scale when present, usually immersed in mucus, may be present on head also 3. Mouth subterminal to inferior 4. Lips thick, fleshy and papillated 5. pairs of barbells 6. An erectile spine before or below eye present 7. Verticle fins with out any spine 8. Pelvic fins may be absent 9. Two sub families with eight genera	**Subfamily:** Botinae 1. Two pairs of rostral barbells placed very close to each other 2. Caudal fin well forked 3. Dorsal fin with 10 to 15 rays (8 to 13 branched)	Botia	*1.* *Botia* spp – with 11 species

Family	Key Characters	Subfamily	Genus	Species
		ub family – Cobitinae 1. 1. Only one pair of rostral barbells. Caudal fin rounded (or) Slightly emarginated (exception Acantopsis) 2. 2. Dorsal fin with 8 – 30 rays	7 Genera	1. . *Neoeucirrhichthys mayelli* - single speices 2. *Enobarbichthys macularus* - single species 3. *Pangio* spp - 4 species 4. *Somileptus gongota* - single species 5. *Lepido cephalichthys* sp. - 11 species 6. *Misgurnus anguillicaudatus* - only one species 7. *Acantopsis* sp - two species

Family	Key Characters	Subfamily	Genus	Species
Balitoridae	1. Body covered with small cycloid scales 2. Depressed ventral surface, some without scales 3. Gill openings restricted or moderate, united membranes 4. Absence of pseudobranchiae 5. Pharyngeal teeth vary, some with an adipose ridge 6. Paired fins may or may not be horizontally placed	**Subfamily** - Balitori-dae	Bhavania Homaloptera Travancoria Balitora	1. *B. austra-lis* 2. *Homalop-tera* spp (seven spe-cies) 3. *Travan-coria* spp (two spe-cies) 4. *Balitora* spp (three species)
	7. Complete lateral line 8. Elongated operculum, reduced preoperculum 9. Reduced, divided air bladder in bony capsule **Subfamily** - Cobitinae 1. One pair of rostral barbels, rounded caudal fin 2. Dorsal fin with 8–30 rays	**Subfamily** – Nemacheilinae	Acanthocobitis Neemacheilus Neonoemacheilus Schistura Longischistura Physoschistura Mesonoemacheilus Nemachilichthys	1. *Acan-thocobitis* sp (six spe-cies) 2. *Neema-cheilus* sp (four spe-cies) 3. *Neonoe-macheilus* sp (three species) 4. *Schistura* sp (63 spe-cies, includ-ing 5 sub-species) 5. *Longis-chistura* sp (two spe-cies)

Family	Key Characters	Subfamily	Genus	Species
			Triplophysa Oreonectes Aborichthys Yunnanilus	6. *Phy-soschistura* sp (five spe-cies) 7. *Meso-noemachei-lus* sp (nine species) 8. *Nema-chilichthys* sp (two spe-cies) 9. *Triplophysa* sp (17 spe-cies) 10. *Ore-onectes* sp (two spe-cies) 11. *Aborichthys* sp (four species) 12. *Yun-nanilus brevis* (one species)

Chapter-23

Order: Siluriformes (Fresh Water)

Fresh Water Fishes

Order : Siluriformes

Kingdom: Animalia

Phylum: Chordata

Class: Actinopterygii (ray-finned fishes)

Order: Siluriformes

Family: Ictaluridae, Bagridae, Siluridae, Pangasidae, Sisoridae, Claridae, Malapteuridae, Loricaridae, Heteropenustidae, Amblycipitidae, Akysidae and Schilbidae.

Siluriformes Overview:

Total Families: 34

Total Genera: 412

Total Species: 2,405

Available Species: Approximately 1,440

Representation: Siluriformes represent about 32% of all freshwater fishes.

Siluriformes in the Indian Region:

The Indian region, encompassing India, Bangladesh, Nepal, Myanmar, and Sri Lanka, hosts a diverse array of siluroids. In this vast region, there are 13 families distributed among 52 genera, comprising a total of 197 species.

Sl. No.	Family	Genera	Number of Species
1	Bagridae	8	48
2	Siluridae	4	13
3	Schilbidae	6	16
4	Pangasiidae	1	2
5	Amblycipitidae	1	7
6	Akysidae	1	4
7	Sisoridae	15	68
8	Erethistidae	6	14
9	Clariidae	2	3
10	Heteropneustidae	1	2
11	Chacidae	1	2
12	Plotosidae	1	3
13	Ariidae	5	15
Total		52	197

Family	Key Characteristics	Genera	Number of Species
1. Ictaluridae (American catfishes)	1. North American freshwater catfishes 2. Freshwater; North America (Southern Canada to Guatemala) 3. Four pairs of barbels on head; skin naked. 4. Dorsal and pectoral fins with a spine 5. Dorsal fin usually with six soft rays 	Ictalurus, Trigloglanis, Ameiurus, Noturus, Pritella, Saran, Pylodictis.	45

Family	Key Characteristics	Genera	Number of Species
2. Bagridae (Bagrid catfishes)	1. Freshwater, distributed in Africa and Asia (to Japan & Borneo) 2. Dorsal fin has one spine along with 6 or 7 soft rays 3. Adipose fin present and highly variable in size between species 	Austroglanis, Bagarius, Bagrichthys, Bagroides, Chrysichthys, Clarotes, Gagata, Glyptothorax, Hara, Nangra, Pseudolaguvia, Rita, Sperata, Batasio, Olyra, Pelteobagrus, Platyglanis, Pseudobagrus	Approximately 210
3. Siluridae (Sheat catfishes)	1. Freshwater, distributed in Europe and Asia 2. Dorsal fin usually with fewer than seven rays, sometimes absent, not preceded by a spine 3. Anal fin base very elongate (41 – 110 rays) 	Belodontichthys, Ceratoptilichthys, Hemisilurus, Hito, Kryptopterus, Ompok, Parasilurus, Pterocryptis, Silurichthys, Silurus, Wallago	About 100
4. Pangasiidae (Shark catfishes)	1. Freshwater, distributed in southern Asia (Pakistan to Borneo) 2. Usually two pairs of barbells, body compressed 3. Adipose fin present, small and never confluent with caudal fin 	Helicophagus, Pangasius, Pangasianodon	Not specified

Family	Key Characteristics	Genera	Number of Species
5. Sisoridae (Sisorid catfishes)	1. Freshwater, distributed in Southern Asia 2. Body usually with small unculiferous tubercles 3. Thick leathery skin 4. Adipose fin present 5. Dorsal fin base short 6. Four pairs of barbells 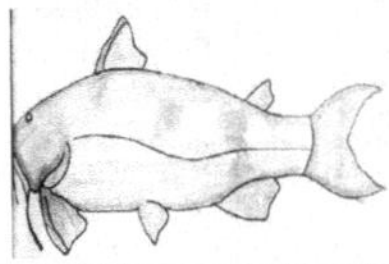	Bagarius, Erethistes, Exostoma, Pseudecheneis, Glyptosternum, Glyptothorax, Hara, Nangra, Puchiloglarus, Sisor	About 85
6. Clariidae (Air breathing catfishes)	1. Freshwater fishes; distributed through Africa, Syria, southern and western Asia 2. Dorsal fin base very long, with more than 30 rays 3. Pectoral and pelvic fins absent in some species 4. Four pairs of barbels present. 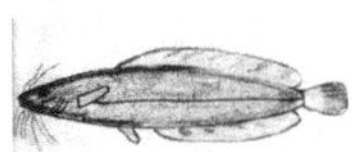	Clariallabes, Clarias, Dinotopterus, Heterobranchus, Horabagrus, Vertiglarius	About 100
7. Malapteruridae (Electric Catfishes)	1. Freshwater, distributed in Africa and Nile 2. Well developed Electrogenic organ present 3. Dorsal fin absent; Adipose fin far back 4. Caudal fin rounded 5. 3 pairs of barbells (nasal pair absent) 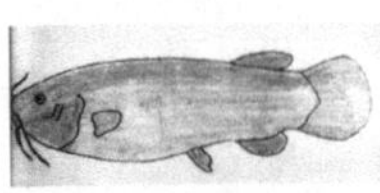	Malapterurus	2

Family	Key Characteristics	Genera	Number of Species
8. Loricariidae (Armored Catfish)	1. Freshwater; distributed in Panama and South America 2. Body with bony plates 3. Mouth ventral, with or without noticeable barbells 4. Ventral lip papillose 5. If adipose fin present, usually with a spine 	Multiple subfamilies with 80 genera	550
9. Heteropneustidae (Air sac catfish)	1. Commonly known as Sting fishes, also air-breathing with tubular air sacs 2. Two known species: *H. fossilis* and *H. longipectoralis* 3. Barbels 4 pairs. 	Heteropneustes	2
10. Amblycipitidae (Torrent catfishes)	1. Characterized by the presence of a fold of skin above anterior to pectoral fin base 2. Single genus Amblyceps with ten species 3. 4 pairs of barbels present 	Amblyceps	10

Family	Key Characteristics	Genera	Number of Species
11. Akysidae (Stream catfishes)	1. Pectoral fins placed horizontally with a strong osseous spine, furrowed longitudinally, serrated or smooth, covered with thick skin 2. A single genus with four species 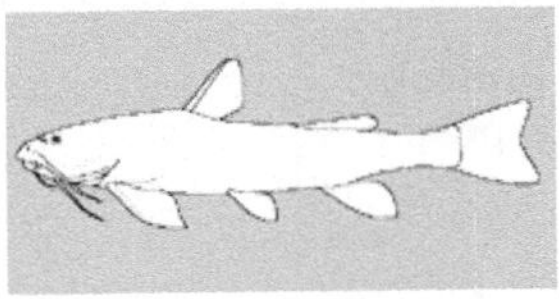	Akysis	4
12. Schilbidae (Schilbid catfishes)	1. Widely separated nostrils, along the border of snout, posterior one slit-like with a flap 2. Barbels elongated, one, two, or four pairs 3. Anal fin very long, free from caudal fin 4. 4 Pairs of barbels present. 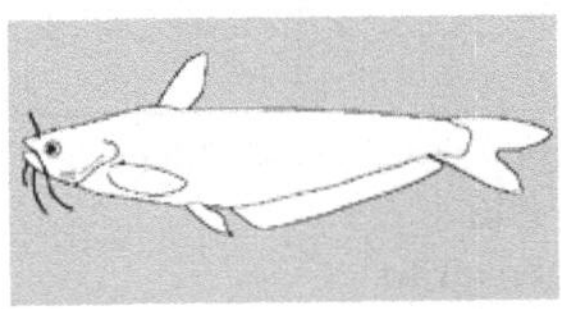	Multiple subfamilies with about 16 genera	40

Barbels

- **Wallago attu** – 2 pairs
- **Pangasius** – 2 pairs
- **Ompak** – 2 pairs
- **Arridae** – 3 pairs
- **Rita** – 3 pairs
- **Tachysuridae** – 3 pairs
- **Malapteruridae** – 3 pairs
- **Claridae (clarius batrachus /magur)** – 4 pairs
- **Mystus** – 4 pairs
- **Sisoridae** – 4 pairs

- **Heteropneustes fossilis** – 4 pairs
- **Bagarius** – 4 pairs
- **Plotossidae** – 4 pairs
- **Mullidae** – 2 barbels on chin

Commercially Important Catfish Species Found in India

Common Name	Scientific Name
Pangas or Pangas Catfish	*Pangasius pangasius*
African Catfish or African Sharptooth Catfish	*Clarias gariepinus*
Magur or Walking Catfish	*Clarias magur*
Stinging Catfish or Singhi	*Heteropneustes fossilis*
Butter Catfish or Baim	*Ompok bimaculatus*
Gangetic Mystus or Tengra	*Mystus cavasius*
Helical Walking Catfish or Boal	*Wallago attu*
Striped Eel Catfish or Boal-Magur Hybrid	*Sperata aor*
Pangasius Catfish or Pangas	*Pangasianodon hypophthalmus*
Gangetic Sea Catfish or Boari	*Clupisoma garua*
Giant Devil Catfish or Indian Goonch	*Bagarius yarrelli*
Sun Catfish or Hora	*Horabagrus brachysoma*
Yellow Catfish or Jaal Magur	*Mystus gulio*
Burmese Bagrid Catfish	*Hemibagrus wyckioides*
Pabo Catfish or Butter Catfish	*Ompok pabo*
Walking Catfish or Magur	*Clarias batrachus*
Indian Dwarf Catfish or Tengara	*Mystus tengara*
Aor Catfish or Aor Magur	*Aorichthys aor*

Chapter-24

Order: Channiformes

Fresh Water Fishes

Order: Channiformes

Kingdom: Animalia

Phylum: Chordata

Class: Actinopterygii (ray-finned fishes)

Order: Channiformes

Family: Channidae

Body Structure

- Body elongate with scales
- Head with plate-like scales

Respiratory System

- Gill openings wide
- Suprabranchial accessory respiratory apparatus well developed

Gill Features

- No pseudobranchiae

Branchiostegals five

Dorsal and Anal Fins

- ❑ Teeth on jaws and palate
- ❑ Single long dorsal fin, and a similar anal fin, both without spines

Pelvic Fins and Lateral Line

- ❑ Pelvic fins thoracic or absent
- ❑ Lateral line abruptly curved or almost interrupted
- ❑ Airbladder present

Taxonomic Features

- ❑ Single family with a single genus (Channidae)

Family - Channidae (Murrels)

Body Morphology

- ❑ Body elongate, subcylindrical anteriorly.
- ❑ Body with large scales; head with plate-like scales

Head and Mouth

- ❑ Cephalic pits present
- ❑ Teeth on jaws, vomer, and palate

Gill Characteristics

- ❑ Branchiostegals five
- ❑ Pseudobranchiae absent
- ❑ Gills four
- ❑ Gill membranes connected beneath isthmus

Respiratory Organ

- ❑ Suprabranchial organ present

Dorsal, Anal, and Caudal Fins

- ❑ Single long spineless dorsal fin present, and a similar anal fin, both free from caudal fin

Lateral Line and Air-bladder

- ❑ Lateral line abruptly curved or almost interrupted
- ❑ Air-bladder present, not bifid posteriorly

Internal Anatomy

- ❑ Pyloric appendages few, when present
- ❑ Basisphenoid roughly Y-shaped

- ❑ Frontals articulating with the basisphenoid, not with the parasphenoid
- ❑ Posterior edge of nasal suturally united with frontal
- ❑ Each epiotic produced into a prominent epiotic process posteriorly

Taxonomic Features

Criteria	Species
1. Scales and Predorsal Scales	
❑ Four or five scales between orbit and angle of preopercle, 12 or 13 predorsal scales	2
❑ Nine to 17 scales between orbit and angle of preopercle, 15 or 22 predorsal scales	4
2. Pelvic Fin and Pectoral Fins	
❑ Pelvic fin less than half length of pectoral fin. Pectoral fins spotted in zones, darker and lighter patches. Chin marbled.	3
❑ Pelvic fin more than half length of pectoral fin. Pectoral fins plain.	*C. punctatus*
3. Anal Fin and Dorsal Fin	
❑ Anal fin with 26 rays. Circular black spots each occupying part of a scale present, base of dorsal iridescent blue in life. Dorsal fin with 39 - 40 rays.	*C. stewartii*
❑ Anal fin with 21-23 rays. No black spots on scales. Dorsal fin with 32-37 rays.	*C. orientalis*
4. Scales and Predorsal Scales	
❑ 16 or 17 scales between orbit and angle of preopercle, 22 predorsal scales	*C. micropeltes*
❑ Nine or 10 scales between orbit and angle of preopercle	5
5. Preopercle Scales and Lateral Line	
❑ 18 – 20 preopercle scales. Lateral line with peninsula of the darker color extending into the yellow.	*C. striatus*
❑ 15-16 preopercle scales. No such color markings on lateral line	6
6. Caudal Fin	
❑ A black white - edged ocellus on basal portion of caudal fin present.	*C. marulius*
❑ No such ocellus	7

Criteria	Species
7. Lateral Line Scales	
❑ Lateral line scales 80	*C. amphibia*
❑ Caudal fin rounded. Lateral line scales 60 – 65.	*C. barca*

Common Murrels in India

Common Name	Scientific Name
Striped Snakehead Murrel	*Channa striata*
Indonesian Snakehead	*Channa micropeltes*
Blotched Snakehead	*Channa maculate*
Great Snakehead	*Channa marulius*
Northern Snakehead	*Channa argus*
Javanese Snakehead	*Channa Lucius*
Emperor Snakehead	*Channa marulioides*
Spotted Snakehead (or Murrel)	*Channa punctata*
Orange spotted Snakehead	*Channa aurantimaculata*
Ocellated Snakehead	*Channa pleurophthalma*
Rainbow Snakehead	*Channa bleheri*
Borna Snakehead	*Channa amphibeus*
Orange finned Snakehead	*Channa aurantipectoralis*
Burmese Snakehead	*Channa harcourtbutleri*
Bangka Snakehead	*Channa bankanensis*
Goonch (Devil Catfish)	*Bagarius yarrelli*

Channidae

Chapter-25

Order: Anabantiformes

Introduction

Anabantiformes, known as labyrinth fish, are freshwater species recognized for their labyrinth organ, enabling them to breathe air. Primarily found in tropical regions like Southeast Asia and Africa, they thrive in oxygen-deprived waters. Popular members of this order include bettas and gouramis, cherished for their vibrant colors and elaborate fins in aquariums worldwide. Anabantiformes exhibit unique reproductive strategies, such as bubble nesting.

Systematic classification

Kingdom: Animalia

Phylum: Chordata

Class: Actinopterygii

Order: Anabantifomes

Family: Anabantidae

Identification Key Characters of Family: Anabantidae (Climbing gouramies)

- Fixed conical teeth on jaws, prevomer, and parasphenoid.
- Relatively large mouth.
- Upper jaw slighlty protractile.
- The genus Sandelia has only cycloid scales, few gill rakers, and generally a carnivorous diet.

Layout

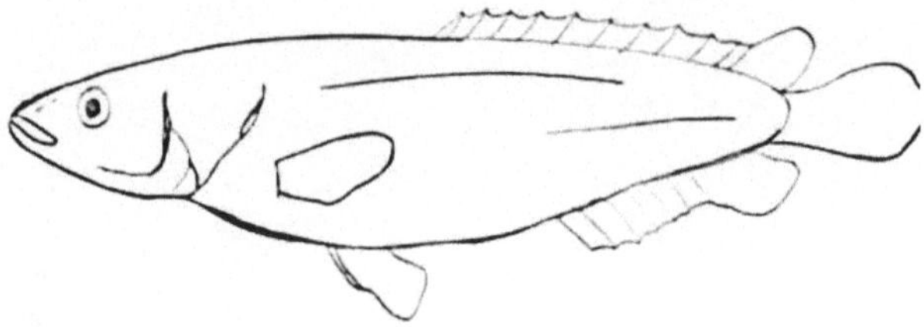

Chapter-26

Order: Cichliformes

Introduction

Cichliformes encompasses a diverse group of freshwater and marine fish, with cichlids being the most well-known family. These fish are admired for their varied colors, behaviors, and breeding strategies, making them popular in aquariums. Typically possessing a compressed body shape and large mouths with specialized teeth, cichlids exhibit complex social behaviors and adaptability to various habitats, including freshwater, brackish, and marine environments.

Systematic classification

Kingdom: Animalia

Phylum: Chordata

Class: Actinopterygii

Order: Cichliformes

Family: Cichlidae and Pholidichthyidae

Identification Key Characters of Family: Cichlidae(Cichlids)

- Body shape quite variable, mostly moderately deep and compressed.
- A nostril on each side of head.
- Interrupted lateral line in most species.
- Scales in lateral lines may be over 100, usually 20-50.
- Dorsal fin usually with 7-25 spines and 5-30 soft rays.
- Spines in anal fin 3-15 (generally 3); soft rays 4-15 (a few with 30).

- Subocular shelf absent.

Layout

Identification Key Characters of Family: Pholidichthyidae(Convict blenny)

- A nostril to each side of head. Scaleless.
- Pelvic fins below or a little before pectoral base.
- One weak spine in pelvic fin; soft rays 2 or 3.
- Caudal fin confluent with anal and dorsal fins.
- Soft rays in dorsal fin 70-79.
- Anal soft rays 55-62.
- Pectoral fin rays 15.
- Lower pharyngeal bones coalesced into one.

Layout

Chapter-27

Mammals

Introduction

The superclass Tetrapoda encompasses a diverse group of vertebrates, all sharing a common evolutionary origin with four limbs or their descended equivalents. Within Tetrapoda, three prominent classes stand out, each representing a distinct branch of the vertebrate tree of life.

Classes under Tetrapoda

- **Class Aves (Birds):** This class includes birds, which are characterized by features such as feathers, beaks, and a lightweight skeleton adapted for flight. Birds are warm-blooded and lay eggs.
- **Class Mammalia (Mammals):** This class includes mammals, which are characterized by mammary glands that produce milk to nourish their young, hair or fur covering their bodies, and a warm-blooded metabolism. Mammals give birth to live young, although there are exceptions, such as monotremes (platypus and echidna) that lay eggs.
- **Class Reptilia (Reptiles):** This class includes reptiles, which are characterized by features such as scales, claws, and usually a cold-blooded metabolism. Reptiles lay amniotic eggs, which have a protective shell, and they are typically adapted to various terrestrial environments.

Mammals

CLASSIFICATION OF MARINE MAMMALS

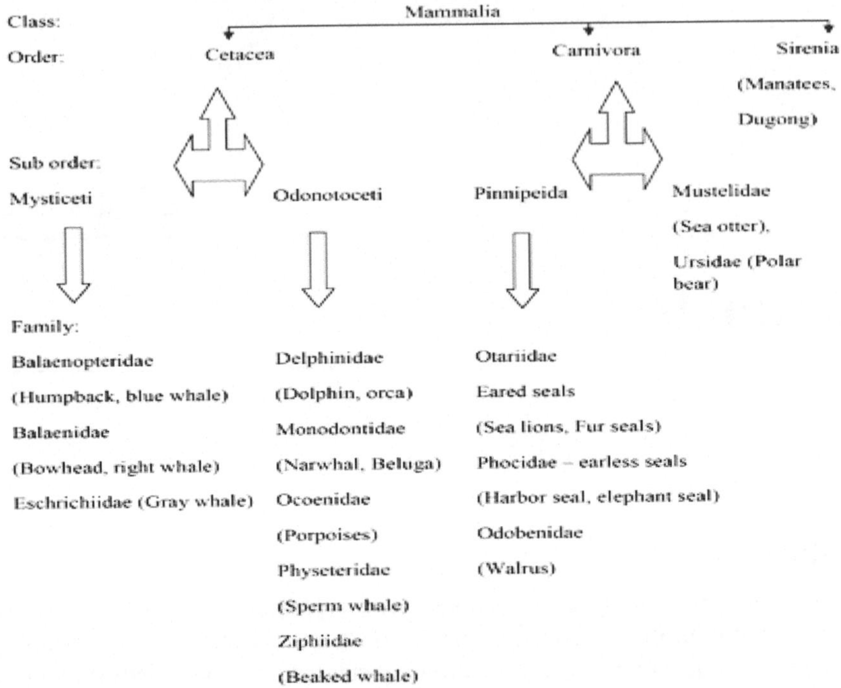

Most Common Group of Mammals

Pinnipeds

Families: Otariidae (sea lions and fur seals), Phocidae (seals), Odobenidae (walruses).

Carnivores

Families: Ursidae (polar bears), Mustelidae (sea otters).

Cetaceans

Suborders: Mysticeti (baleen whales), Odontoceti (toothed whales like sperm whales, orcas, dolphins, and porpoises).

Mysticetes use baleen plates to filter zooplankton and phytoplankton, while Odontocetes have teeth for capturing prey.

Sirenians

Families: Dugongidae (2 species of dugongs), Trichechidae (3 species of manatees).

Cetaceans Overview

- ❑ The term "cetacean" encompasses whales, dolphins, and porpoises in the order Cetacea.
- ❑ Derived from Latin "cetus" (large sea animal) and Greek "ketos" (sea monster).
- ❑ Cetaceans range widely in size, from the small Hector's dolphin to the colossal blue whale.
- ❑ They inhabit oceans and major rivers globally.
- ❑ Thought to have evolved from hooved land mammals around 50 million years ago.

Mysticetes

- ❑ **Examples:** Blue whale, fin whale, right whale, humpback whale.
- ❑ Feed using baleen plates to filter prey from large gulps of water.

Odontocetes

- ❑ Comprise 72 species, including sperm whales, orcas, belugas, dolphins, and porpoises.
- ❑ Have cone-shaped or spade-shaped teeth for capturing prey.
- ❑ Typically feed on fish and squid, with some orcas preying on other marine mammals.
- ❑ Highly developed senses.
- ❑ Presence of blubber for insulation.
- ❑ Long bodies for efficient swimming.
- ❑ Distinct features differentiate baleen whales from toothed whales, such as baleen plates and teeth.

Suborder: Mysticeti

Family	Characteristics	Examples
1. Balaenidae	Large and chunky with heads comprising up to one-third of their body length. 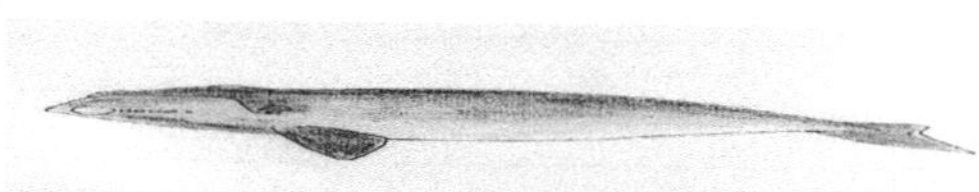	e.g., *Balaena mysticetus* (right whale)
2. Neobalaenidae	Head represents about one-quarter of the total length. - Short falcate dorsal fin set behind midback. - Pair of shallow throat grooves. - Skull moderately arched. 	e.g., *Caperea marginata* (pygmy right whale)

Family	Characteristics	Examples
3. Balaenopteridae	Largest animals ever to live. - Rorquals with streamlined bodies and long pleats extending from snout tip to navel. - Baleen plates of moderate length. - Upper jaw has a relatively flat profile. 	e.g., *Balaenoptera musculus* (blue whale)
4. Eschrichtiidae	Stocky with an arched jaw. - Slow-moving coastal animals. - Shortest and coarsest baleen. - 2 to 5 short throat creases. - Dorsal hump and knuckles along the dorsal surface of the tail stock. 	e.g., *Eschrichtius robustus* (gray whale) s

Suborder: Odontocet

Family	Characteristics	Examples
1. Physeteridae	Low dorsal hump followed by crenulations. Large head with a squarish profile. Narrow underslung lower jaw. Functional teeth only in the lower jaw. Blowhole located at the left front of the head. 	e.g., *Physeter macrocephalus* (sperm whale)
2. Kogiidae	Blunt squarish heads with underslung lower jaws. Smaller heads than in sperm whales. Blowhole not located at the front of the head. Skull structure shares facial area and asymmetry with sperm whales. Larger dorsal fin. 	e.g., *Kogia breviceps* (pygmy sperm whale)
3 .Monodontidae	Small whales (less than 6 m). Stocky bodies, blunt bulbous heads, broad rounded flippers, and no dorsal fins. Flat profile in the skull. Unfused cervical vertebrae for neck flexibility. 	e.g., *Monodon monoceros* (narwhal)

Family	Characteristics	Examples
4. Ziphiidae	Medium-sized cetaceans. Pronounced beak. Small dorsal fin set far back on the body. Small flippers. 2 short throat grooves. 1 or 2 pairs of functional teeth in the lower jaw of males only. 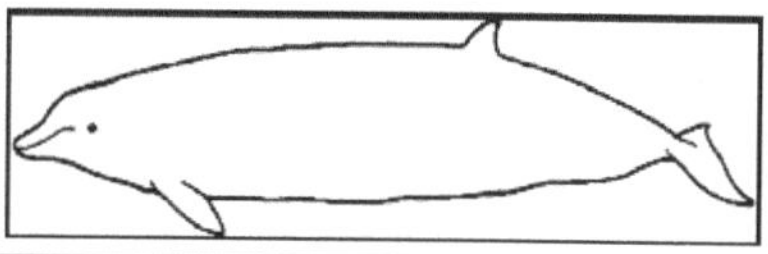	e.g., *Berardius bairdii* (Baird's beaked whale)
5. Delphinidae	- Marine habitat, noticeable beak, conical teeth, and a large falcate dorsal fin set near the middle of the back. 	e.g., *Delphinus delphis* (common dolphin)
6. Phocoenidae	Small cetaceans (all less than 2.5 m). Coastal distribution, rather stocky in form. Short triangular dorsal fin. Spade-shaped teeth. Females are larger in some species. 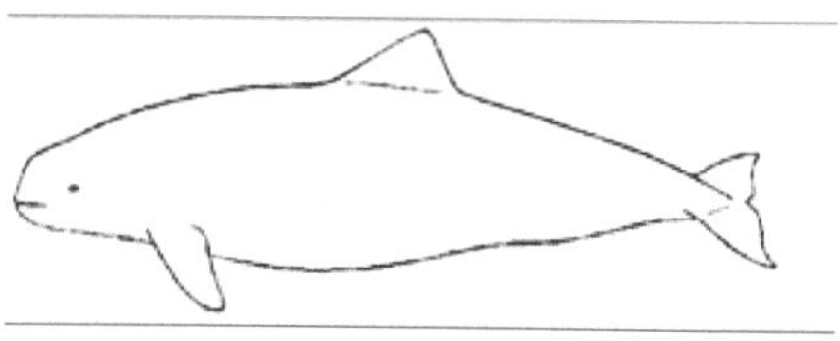	e.g., *Phocoena phocoena* (harbor porpoise)
7. Platanistidae	Susu and bhulan of the Ganges and Indus rivers. Nearly blind, rely on echolocation. - Small "mushy" bodies. - Long foreceps-like beak. - Blowhole is a longitudinal slit. - No true dorsal fin. 	e.g., *Platanista gangetica* (Ganges river dolphin)
8. Iniidae	Boto of the Amazon and Orinoco drainages in South America. Moderately long thick snout with sparse hairs. - Very low and usually indistinct dorsal ridge. Pink coloration in many adults. 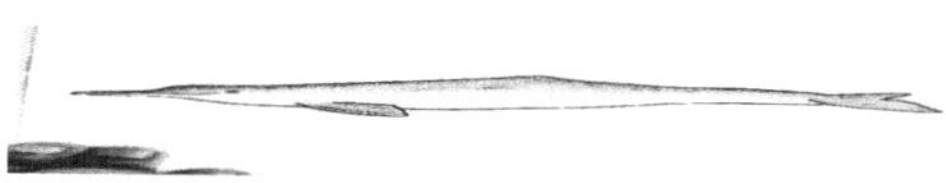	e.g., *Inia geoffrensis* (Amazon river dolphin)
9. Pontoporiidae	Two types of dolphins: one a true river dolphin of the Yangtze River, the other a coastal marine species of the east coast of South America. Long beaks, low triangular dorsal fins. 	e.g., *Pontoporia blainvillei* (La Pslata dolphin)

Pinnipeds

- Three families of pinnipeds
- The Family Otariidae (sea lions and fur seals)
- Family Phocidae (true seals)

General Characteristics of Pinnipeds:

- Pinnipeds, including seals, sea lions, and walruses, have blubber for insulation.
- They are warm-blooded animals.
- Pinnipeds can live both on land and in water.
- Seals, in particular, can spend extended periods at sea, returning to land for mating and molting.
- They are found in coastal areas worldwide.

Characteristics	Otariidae (Eared Seals)	Phocidae (True Seals)	Odobenidae (Walruses)
Front Flippers	Large and capable of rotation.	Short flippers with no rotation.	N/A (Walruses have front flippers.)
Ear Lobes	Tiny ear lobes.	Lack external ears.	N/A (Walruses have small external ears.)
Back Legs	Webbed, capable of rotation.	Non-rotating back flippers.	N/A (Walruses have hind flippers.)
Ears	External ears present.	No external ears.	Small external ears.
Land Movement	Agile movement on land.	Awkward movement on land.	Awkward movement on land.
Hair on Body	Some fur present.	Sparse fur.	Some hair present.
Tusks	No tusks.	No tusks.	Long tusks.
Whiskers	Whiskers present.	Whiskers present.	Short, thick whiskers.
Size	Generally smaller compared to walruses.	Various sizes, generally smaller than walruses.	Generally larger compared to seals and sea lions.
Examples	e.g., **Northern Fur seal** - *Callorhinus ursinus*	e.g., **Harbor seal** - *Phoca vitulina*	e.g., **Walrus** - *Odobenus rosmarus*
Layout			
Ursidae (Polar bear)	- Largest land carnivores.		e.g., *Ursus maritimus* 
	- Found in the Arctic on sea ice, islands, and coastlines.		
	- Feed on ringed, bearded, harp, and hooded seals.		
	- Black skin with dense underfur and transparent guard hairs.		
	- Thick layer of blubber for temperature protection.		
Mustelidae (Sea ottar)	Short legs.		
	Short, round ears.		
	Thick fur.		
	Most mustelids are solitary, nocturnal, and active year-round.		

Sirenia: Order Sirenia

Includes three species of manatees and two species of dugongs.

Manatees: Characteristics

- Have a round tail.
- Found in Southeast U.S. coastal waters, Central America, and northern South America.
- Gentle animals with large front flippers.

Dugongs: Characteristics

- Have a dolphin-like tail.
- Found in coastal waters from East Africa to the South Pacific.
- Higher blubber ratio compared to other marine mammals.
- Gentle animals with large front flippers.

Family	Characteristics	Examples
1. Trichechidae (Mantees)	- Round tail. - Found in Southeast U.S. coastal waters, Central America, and northern South America.	e.g., *Trichechus manatus* 
2. Dugongidae (Dugong)	- Dolphin-like tail. - Found in coastal waters from East Africa to the South Pacific.	e.g., *Dugong dugong* 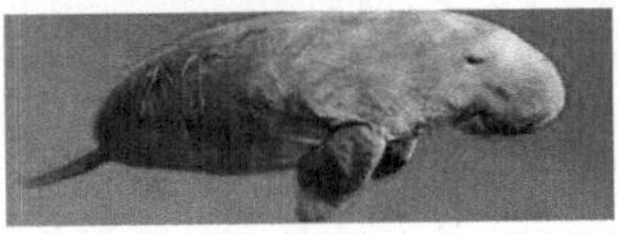

Chapter-28

Reptiles (Turtles)

Characteristics

- Air-breathing, ectothermic (cold-blooded) vertebrates.
- Lay shelled eggs.
- Skin covered in scales and/or scutes.
- Creeping tetrapods, either having four limbs or descended from four-limbed ancestors.
- Evolved around 320-310 million years ago during the Carboniferous period.

Size Range

- Largest reptile: *Crocodylus porosus* (Saltwater Crocodile).
- Smallest reptile: *Sphaerodactylus ariasae* (A Caribbean gecko).

Living Orders:- Order Testudines (Turtles)

- Characterized by a bony or cartilaginous shell.
- Includes both terrestrial and aquatic species.

Order Rhynchocephala (Tuataras)

- Tuataras are lizard-like reptiles found in New Zealand.
- Unique among reptiles with a third eye on the top of their head.

Order Squamata (Snakes, Lizards, and Worm Lizards)

Suborder Lacertilia (Lizards)

- Includes various species of lizards with diverse habitats and characteristics.

Suborder Amphisbaenia (Worm Lizards):

- Legless, burrowing reptiles with a cylindrical body.

Suborder Serpentes (Snakes)

- Legless, elongated reptiles with a highly flexible body.
- Includes a wide range of species adapted to different environments and feeding habits.

Order Crocodilia (Crocodiles)

- Large aquatic reptiles with elongated snouts and powerful jaws.
- Well-adapted to aquatic life, inhabiting freshwater habitats.

General Remarks on Sea Turtles

Historical Background

- Sea turtles date back at least 200 million years.
- Coexisted with dinosaurs and survived various ancient marine reptiles.
- Present-day genera and species originated between the early Eocene to the Pleistocene (60 to 10 million years ago).
- Among the surviving sea-water-adapted reptiles along with marine snakes and iguanas.

Distribution and Dependency

- Mostly tropical and subtropical distribution.
- Depend on land only during the reproduction period.
- Some viviparous sea snakes also share this habitat.

Commercial Importance

- High commercial value globally.
- Varying importance among countries, with different uses such as protein source, delicacy, and cultural significance.
- Commercialization increased significantly post-World War II.

Threats and Conservation

- Rapid depletion of colonies due to commercialization.
- Additional threats include pollution, beach invasion, poaching, entanglement in set-nets, and drowning in trawl-fishing.
- Almost all sea turtles are considered threatened or endangered according to IUCN and CITES.

- Conservation efforts include law enforcement, fishing regulations, gear restrictions, protection of nesting areas, hatchling release, and imprinting-headstarting programs.

Commercial Uses

- Sea turtle catch primarily for fresh meat consumption.
- Egg-harvesting for riparian people in many countries.
- Leather made from turtle skin.
- Turtle oil used in cosmetics and medicine.
- Varies in commercial value among species.

Conservation Programs

- Worldwide actions supported by local and international wildlife foundations and governmental organizations.
- Includes law enforcement, fishing regulations, gear restrictions, protection of nesting areas, and hatchling release.

Augmenting Populations

- Ranching involves raising sea turtles from eggs or hatchlings taken from the wild to a marketable size.
- Farming relies initially on wild populations for eggs or hatchlings but aims for a self-sustaining captive population.

Controversies and Regulations

- Controversy surrounds ranching and farming practices.
- CITES regulates commercialization, allowing trade with captive-bred animals under specific conditions.

Catch Statistics

- Official catch statistics are limited.
- Reported world catch of adults in 1987 totaled 3,100 metric tons, with varying contributions from different fishing areas.

Systamatic Classification

Kingdom: Animalia

Phylum: Chordata

Subphylum: Vertebrata

Class: Reptilia

Order: Testudines

Family: Cheloniidae and Dermochelyidae.

Subclass: Anapsida

They are primitive reptiles with a solid skull roof

Order – Testudines

This order includes all turtles and tortoises. It is divided into three suborders:

- Pleurodira (side-necked turtles)
- Cryptodira (freshwater turtles, snapping turtles, tortoises, soft-shelled turtles and sea turtles)
- Amphichelydia (a suborder of turtles that is now extinct)

Families

Most scientists recognize seven species of sea turtles under two families:

1. Cheloniidae (Hard-shelled sea turtles): This family includes five of the seven species of sea turtles.

Carapace Shape: The upper shell, or carapace, is relatively flat compared to other turtles and is often described as oval or heart-shaped.

Plastron Size and Attachment: The lower shell, or plastron, is somewhat smaller than in many other turtles and attaches to the upper shell through tough but flexible tissues called ligaments.

Modified Limbs: The limbs of sea turtles are adapted into flippers, which are specialized for swimming rather than walking on land.

Inability to Retract Limbs or Head: Unlike many other turtles, sea turtles cannot retract or pull back their limbs or heads into their shells. This characteristic distinguishes them from other turtle species.

Cheloniidae Family: The family Cheloniidae includes six species under six genera. These species are generally large, with varying maximum carapace lengths, as mentioned (ranging between 71 cm for Lepidochelys olivacea and 213 cm for Caretta caretta).

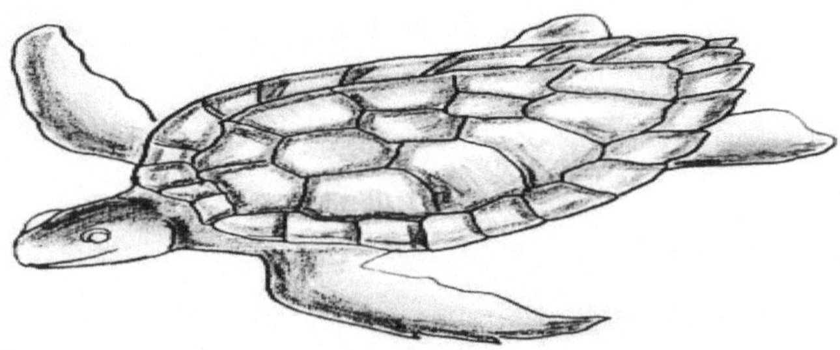

- **Loggerhead sea turtle** (*Caretta caretta*)

Large head with a reddish-brown shell, powerful jaw for crushing prey (mainly crustaceans), and a distinctive heart-shaped carapace.\

- **Australian Flatback** (*Natator depressus*):

Named for its flat back. Found only in the waters of Australia. Can weigh up to 200 pounds, reach 40 inches in length.

- **Green sea turtle** (*Chelonia mydas*)

Greenish color of their fat and cartilage, herbivorous diet (mainly seagrasses and algae), and a smooth, heart-shaped carapace.

- **Hawksbill sea turtle** (*Eretmochelys imbricata*)

Sharp, pointed beak resembling a hawk's bill, a unique overlapping shell pattern, and primarily feeds on sponges and other invertebrates on coral reefs.

- **Kemp's ridley sea turtle** (*Lepidochelys kempii*)

Olive-gray coloration, a relatively small and round carapace, and primarily found in shallow coastal waters.

- **Olive ridley sea turtle** (*Lepidochelys olivacea*)

Olive-colored carapace, a more rounded and heart-shaped shell, and known for mass nesting events called arribadas where females come ashore together to lay eggs.

2. Dermochelyidae (Leatherback sea turtle): This family consists of one species.

- Scuteless turtles with only one modern species;
- The leatherback turtle - covered with leathery skin
- It is the only marine turtle whose backbone is not attached to the inside of its shell

- **Species:** Leatherback (*Dermochelys coriacea*) - the largest living turtle
- Fresh water chelonians called terrapins. Trionyx is a soft shelled fresh water edible terrapin

- **Leatherback sea turtle (*Dermochelys coriacea*)**

The Leatherback sea turtle *(Dermochelys coriacea)* is distinguished as the largest living sea turtle, boasting an average length of six feet and a remarkable weight of up to 1,400 pounds. What sets this species apart is its unique physical features, notably the absence of a traditional shell. Instead, the Leatherback sports a distinctive leathery back adorned with raised gray stripes. This exceptional combination of size and distinct characteristics contributes to the extraordinary nature of the Leatherback sea turtle.

Chapter-29

Reptiles (Snakes and Lizards)

Classification

Kingdom : Animalia

Phylum : Chordata

Subphylum : Vertebrata

Class : Reptilia

Order : Squamata

Suborder : Serpentes

Superfamily:Xenophidia (Colubrids, vipers, and relatives)

Family : Hydrophiidae(sea snakes)

Order Squamata(Overview)

- Includes lizards and snakes
- Most successful living reptiles
- Skull has lost both temporal vacuities

Common Characteristics:

- Cloacal aperture is transverse
- Single-headed ribs
- Movable quadrate
- Vertebrae are procoelous

Characters of True Sea Snakes

Physical Features:

- Flattened body
- Short oar-like tail
- Valvular nostrils on the top of the snout

Internal Adaptations:

- Elongated lungs that extend the entire body length
- Very small scales, not overlapping, abutting against one another
- Reduced belly scales in primitive species; absent in advanced forms

Family: Elapidae - Subfamily: Hydrophiinae	
Aquatic Adaptations	❑ Advanced species unable to crawl on land ❑ Keel formed along part of the belly when swimming, aiding propulsion ❑ Can remain submerged for several hours, possibly up to eight ❑ Breathe through their skin 

Family: Colubridae (Freshwater Snakes)	
Common Species	❑ *Xenochrophis piscator* - Non venomous freshwater snake found in India
Identification Features	❑ Uniform chessboard pattern of black spots on its body ❑ Two tear marks just below the eyes
Diet	❑ Feeds mainly on small fishes and water frogs

Layout	

Lizards within Order Squamata – Families	
1. Iguanidae	❑ **Example:** *Iguana iguana* (Green Iguana) ❑ Often found in tropical regions, known for their distinctive dewlap 
2. Gekkonidae	❑ **Example:** *Hemidactylus frenatus* (Common House Gecko) ❑ Adaptable to various environments, including urban areas 
3. Agamidae	❑ **Example:** *Pogona vitticeps* (Bearded Dragon) ❑ Popular as pets, known for their distinctive appearance and behavior 

Chapter-30

Modern Taxonomic Tools

Why Modern Taxonomic tools?

Earth is a vibrant planet, teeming with millions of diverse species that collectively form the intricate web of life. However, discerning and categorizing this vast array of life forms is a formidable challenge. Taxonomists, equipped with their expertise, can only critically identify a mere 0.01% of the estimated 10–15 million species inhabiting our planet. The staggering diversity of life on Earth demands a fast and reliable classification system, underscoring the need for innovative approaches and technology to unravel the mysteries of the countless species that share our planet.

Limitations of Classical Taxonomic Tools or Morphological Identification

Phenotypic Plasticity and Genetic Variability

One major limitation of classical taxonomic tools lies in the presence of phenotypic plasticity and genetic variability within the characters used for species recognition. This variability can result in incorrect identifications, as the morphological traits may not reliably represent the true genetic diversity among species.

Morphologically Cryptic Taxa

Classical taxonomic methods often overlook morphologically cryptic taxa, commonly found in various groups, including fishes. These hidden species lack discernible phenotypic differences, complicating their identification and highlighting the need for more advanced techniques beyond morphology alone.

Limited Applicability across Life Stages and Genders

Morphological keys frequently prove effective only for a specific life stage or gender, posing a challenge in identifying many individuals correctly. This limitation hinders comprehensive species identification and may lead to misidentification, particularly in cases where morphological traits vary across different developmental stages or between genders.

Expertise Requirements

The use of classical taxonomic keys demands a high level of expertise, limiting their accessibility to a select group of specialists. This exclusivity can impede the broader scientific community's ability to accurately identify and categorize species, emphasizing the necessity for more user-friendly and inclusive identification tools in taxonomy.

Modern Taxonomic Tools

1. Karyotaxonomy (Species Identification by Karyotyping)

Chromosome

Pioneering Work of Karl Wilhelm von Nageli (1842)

The inception of species identification through karyotyping can be traced back to the observations of Swiss botanist Karl Wilhelm von Nageli in 1842. Von Nageli's keen eye led to the description of thread-like structures within plant cell nuclei, initially termed "transitory cytoblasts," which laid the foundation for the understanding of chromosomes.

Waldeyer's Coined Term "Chromosome" (1888)

In 1888, the term "chromosome" was officially coined by Waldeyer. This nomenclature came about following the development of staining techniques that greatly improved the visibility of these genetic structures. The term itself, derived from the Greek words "chromos" for color and "soma" for body, aptly captures the stained and colorful appearance of chromosomes.

Chromosomes: Architectures of Genetic Information

Genetic Material Packaging

Chromosomes play a pivotal role in species identification as carriers of genetic material. DNA, the blueprint of life, is tightly coiled around proteins known as histones. This intricate packaging allows for the efficient storage and transmission of genetic information.

Karyotyping: A Powerful Tool for Species Identification

The technique of karyotyping, rooted in the historical discoveries of von Nageli and Waldeyer, has become an indispensable tool in species identification. By examining the unique chromosomal patterns of organisms, scientists can unravel the genetic signatures that distinguish one species from another.

Chromosome Classification by Centromeric Position: A Tabular Insight

Types of Chromosomes

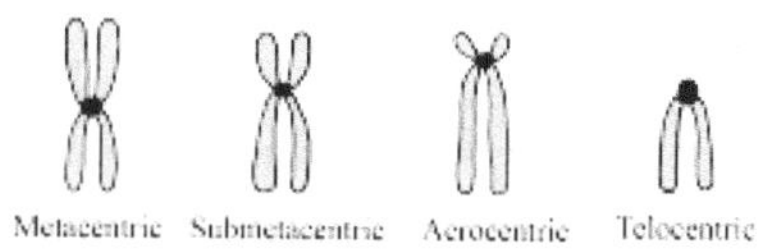

Type	Centromeric Position	Symbol	Centromeric Index Range	Additional Information
Metacentric	Nearly Median	M	46-49	In metacentric chromosomes, the centromere is positioned in the middle, creating symmetrical arms for stable spindle fiber attachment during cell division.
Submetacentric	Submedian	Sm	26-49	Submetacentric chromosomes exhibit an off-center centromere, leading to arms of uneven length and introducing asymmetry in their behavior during cell division.
Acrocentric	Subterminal	St	15-30	Chromosomes with a subtelocentric centromere have this pivotal region close to the terminal end, resulting in one notably shorter arm and influencing behavior during genetic material separation.
Telocentric	Terminal	T	<15	Telocentric chromosomes feature a centromere located at the terminal end, and their unique configuration may impact chromosome movement dynamics during cell division. The entire chromosome is involved in spindle fiber attachment.

Chromosome Parameters: Insights into Structure and Genome Representation

Paramaeter	Formula	Description
Centromeric Index	Length of the short arm ------------------------------ Length of the long arm	Provides a percentage representation of how much the short arm contributes to the entire chromosome.
Arm Ratio	Length of the long arm ------------------------------ Length of the short arm	An alternative method to the centromeric index, always greater than or equal to one, describing the position of the centromere and arm lengths.
Relative length	Length of a chromosome×100 ------------------------- Total length of all chromosomes in the haploid set	Indicates the proportion of the whole genome represented by a specific chromosome. A linear measurement, not reflective of mass and volume.

Understanding Karyotype: An Overview

Karyotype

Referring to the chromosome complement of a cell or an entire organism, a karyotype provides crucial insights into the number, size, and shape of chromosomes during metaphase of mitosis.

Karyotyping

A karyotype is an organized profile of an individual's chromosomes, presented as a pictorial display during metaphase. The process involves staining chromosomes to reveal their structure, aiding in the arrangement of homologous chromosomes in pairs. The chromosomes are then numbered and organized based on the position of the centromere, from largest to smallest. If sex chromosomes are present, they are typically placed last.

Preparing Karyotype

Metaphase cells, essential for standard karyotyping, can be derived from various dividing cell populations such as blood cells or kidney cells. The preparation involves using mitotic inhibitors like colchicine, a hypotonic solution for cell swelling, a fixative like Cornay's fixative, and staining with Giemsa.

Karyotype Formula

The karyotype formula involves categorizing chromosomes based on their arm ratio (q/p) into metacentric (m), submetacentric (sm), acrocentric (st), and telocentric (t). The overall formula is represented as m + sm + st + t, and the Fundamental Arm Number (FN) is calculated as 2xM + 2xSM + ST + T.

Sex Chromosomes in Fishes

Fish exhibit various sex chromosome mechanisms, including male heterogamety (XX-XY) and female heterogamety (ZZZW). Multiple sex chromosomes are also observed in certain species. Techniques such as banding and molecular approaches with sex-specific DNA probes aid in identifying and understanding sex determination systems in fishes.

Diploid Chromosome Numbers in Commercially Important Species

Diploid chromosome numbers vary among commercially important fish species, providing essential data for karyotaxonomy. Some examples include Salmo salar (58 chromosomes), Carassius auratus (104 chromosomes), and Penaeus monodon (88 chromosomes).

Sl. No.	Species	Chromosome No.
1.	*C. punctatus*	16
2.	*Gonostoma bathyphylum*	12
3.	*Channa striatus*	40
4.	*Grass carp*	48
5.	*Silver carp*	48
6.	*Catla, Rohu, Mrigal*	50
7.	*Clarias batrachus*	50
8.	*Oreochromis mossambicus*	44
9.	*O. niloticus*	44
10.	*Heteropneustes fossilis*	58
11.	*Salmo salar*	58
12.	*C. gachua*	78
13.	*Penaeus monodon*	88
14.	*Carassius auratus*	104
15.	*Cyprinus carpio*	104

Next Level of Karyotaxonomy

Advancements in karyotaxonomy involve techniques like CMA3 staining for constitutive heterochromatin, AgNOR staining for nucleolar organizer regions, and FISH (Fluorescence in situ hybridization) for nucleic acid hybridization, contributing to a deeper understanding of chromosomal structures and functions.

Cytotaxonomy in Fishes

Cytotaxonomy of Finfishes: Insights into Chromosomal Analysis

Karyotype analysis, a fundamental aspect of cytogenetics, involves studying the chromosome number, size, and morphology during the metaphase stage. The resulting karyotypes, which are heritable characteristics of each species, play a crucial role in taxonomic differentiations, leading to the field known as Cytotaxonomy. Chromosomal rearrangements drive karyotypic variations in different evolutionary lineages, contributing to the field of cytosystematics. Fish cytogenetics, starting in the 1960s, has significantly progressed, aided by advanced techniques and methodologies.

Cytological Techniques

Chromosome Preparation

- In vivo method involves colchicine treatment of actively dividing tissues like kidney, spleen, or gill.
- In vitro method includes short-term culture of blood cells, abdominal fluid, fin tips, or scale epithelium.
- A hypotonic solution is used to swell cells, followed by fixation in methanol:acetic acid (3:1).
- Slides are prepared by flame drying or air drying.

Staining of Chromosomes

- Staining with 4-5% Giemsa solution facilitates visualization.
- Fish karyotypes, often challenging due to numerous small-sized chromosomes, benefit from advancements like chromosome banding techniques.

Chromosome Banding Techniques

Chromosome bands are produced due to the variations in longitudinal structure of the chromatids, revealed by various techniques. Q-banding is named after quinacrine mustard, which is a flurchrome that stains the chromosome differentially. C-band stains deeply the centromeric heterochromatic region present in the centromeric zone of a chromosome. Serial bands (popularly called as G- and R- bands) could be seen throughout the chromosomes by treating them with hot saline solution and staining with Giemsa. Techniques are available to demonstrate nuclear organizer region (NOR), where 18S and 28S rRNA genes are clustered. In last few years the radiolabelled DNA probes are replaced by fluorescent lebelled DNA probes for Fluorescent in situ hybridization (FISH) study. This is useful in detecting the chromosomal location of a DNA sequence.

Nor-banding

Nuclear organizer region on the chromosomes can be detected by Silver staining the metaphase chromosomes. Silver staining of NOR band is relatively a simple method. The mechanism of silver staining is not fully known yet but it is believed that silver stain binds to the non-histone proteins of the transcriptionally active NOR region. The number and location of NOR bands are useful for systematic purpose. In some species it is found that the number of NOR bands is related to ploidy level. For example, in certain species belonging to the Order Acipenseriformes, the increase in ploidy is also accompanied by increase in the number of NOR bands. Acipenser ruthenus (2n=118+4) possesses two pairs of chromosomes. But in the related species A. baceri (2n=250+8) and A. naccani (2n=246+8), four pairs of NOR bands are present, which suggest that these species are tetraploid. However, this is not universal. In some species like Salvelinus NOR bands are found in multiple chromosome location. Interestingly, sex-specific differences in NOR location was observed on the metaphase chromosomes of Mosquito fish, Aplocheilus panchhax.

C-banding

C- band can be demonstrated by

BSG (barium hydroxide/saline/Giemsa) technique developed by Sumner (1972) is regarded as the standard C-banding method. Slides are hydrolysed by 0.2N HCl, denatured in barium hydroxide solution, treated with 2×SSC (saline sodium citrate) and stained with Giemsa.

By staining the metaphase chromosome with restriction enzymes, C- band can also be demonstrated.

Information on C-banding is available in over 100 species of fishes. C-banding in fishes includes not only the centromeric region but often it demonstrates heterochromatin at the telomeric and inter-calary region. In some cases the whole arm is found to be C-band positive.

Fluorescence in Situ Hybridization (FISH)

The radiolabelled DNA probe for in situ hybridization has been replaced by flurocent dye-labelled DNA probe in the recent past. The flurocence signal is detected by the flurocent microscope and various colour patterns on chromosome can be obtained. This technique is now used for homologous chromosome identification and studying the location of a DNA sequence on the chromosome. In the near future greater use of this technique would be found in fish cytotaxonomy and genetic studies.

Applications and Significance

- Chromosomal analysis aids in ploidy determination, species and hybrid identification, and genotoxicity studies related to pollutants.
- NOR banding reveals transcriptionally active regions and is useful for systematic purposes.
- C-banding provides information on centromeric and other heterochromatin regions.

Future Perspectives

Continued advancements in techniques like FISH are expected to enhance fish cytotaxonomy and genetic studies.

DNA Barcoding

DNA Barcoding in Fish Species

Introduction

- DNA barcoding in fish species involves utilizing short, standardized gene regions as "barcodes" to distinguish and identify different fish species, similar to how supermarket scanners use Universal Product Codes (UPC). This method provides a rapid and accurate system for species identification.
- The DNA barcoding is a large scale DNA sequence generation of a uniform DNA fragment for all the species. It is an emerging standard for identifying species and study global biodiversity using gene sequences.
- Mitochondrial gene cytochrome c oxidase subunit I (COI) is the most suitable candidate gene for barcoding and it is being used as a global bioidentification system for animals, including fish.
- Accurate and unambiguous identification of fish and fish products ranging from egg to adults is an indispensable effort for strengthening fish germplasm conservation as well as managing sustainable fisheries.
- DNA barcoding has made it possible to cut the time to identify a species from days to minutes.
- DNA barcode is a tool for taxon recognition and classification of biological organisms based on 650 base-pair sequence of a fragment of mitochondrial gene, cytochrome c oxidase I (COI)
- Useful for species identification, provide support to taxonomist and molecular genetic analysis.
- Over 1700 fish species from india have been barcoded so far.
- DNA barcoding began in 2003 with the proposal that the organisms could be assigned to their correct species using a short gene sequence from a standardized position in the genome.
- In 2005, Consortium for the Barcode of Life (CBOL, www.barcoding.si.edu) launched the Fish Barcode of Life campaign (FISH-BOL; www.fishbol.org) to create a global reference library of all 30,000+ species of cartilaginous and bony fishes from marine, estuarine and freshwater ecosystems, one that is derived from voucher specimens with authoritative taxonomic identifications.
- The benefits include facilitating species identification for all potential users, including taxonomists.
- Mitochondrial genes used for characterization and evolutionary relationship studies

Purpose	Gene/ region
Species Characterization	COI 16S rRNA
Population Characterization	Cytochrome b ATPase8/6

Characteristics of barcode regions

- Flanked by conserved regions
- Balance of variabile and conserved
- Low intraspecies variability
- Long enough to work in all groups
- Short enough for single reads

Cytochrome c oxidase sub unit-I

- Cytochrome c oxidase gene could serve as the basis for a global identification system for animals (Hebert et al., 2003)
- COI gene is the terminal enzyme of the respiratory chain of mitochondria.
- Total length of cytochrome c oxidase I gene in vertebrates is about 1545 base pairs (bp), and 655 bp long commencing near the start of the COI reading frame was selected as the 'barcode' region.
- The hypothesis is that, for COI gene, every species wil have a unique sequence or a unique assemblage of closely related sequences.
- This sequence is termed a 'barcode.

Methodology for DNA Barcoding

- Tissue sampling
- DNA extraction
- PCR amplification
- Sequencing reaction
- Sequencing
- Conversion of sequence to Barcode

DNA Fingerprinting in Fish

While DNA barcoding focuses on species identification, DNA fingerprinting, or DNA profiling, is employed for characterizing the DNA of individual fish. It is applied in various fields, including criminal investigations, paternity testing, immigration status, genealogy, medical research, and the study of animal and plant populations in zoology, botany, and agriculture.

Creating a DNA Barcode Database for Fish

To establish a DNA barcode database for fish species, several essential steps are undertaken.

Sample Collection

Samples are obtained from various sources, including amateur taxonomists' collections and museum collections.

Museum specimens undergo careful screening for reliable DNA quality.

Taxonomic Identification

Taxonomic identification and collection data, including location, date, collector name, etc., are recorded for each sample.

Organisms are photographed in their natural habitat and in the laboratory.

Voucher Storage

Vouchers are stored in the collections of respective institutions participating in DNA barcoding projects.

Laboratory Analysis

DNA extraction is performed by separating a small tissue piece from each sample.

PCR (polymerase chain reaction) amplification replicates DNA barcode regions.

DNA Barcode Generation

A DNA barcode is generated, typically from the cytochrome c oxidase I gene (COI) in mitochondrial DNA for fish.

Quality Check

Professional scientists check sample sequences and metadata for integrity and quality.

Database Upload

Data, including sequences, metadata, classifications, and photographs, are uploaded to international reference libraries like the Barcode of Life Data Systems (BOLD).

Applications in Fish Species

Application	Description
Invasive Species Detection	Effective for detecting and distinguishing invasive species at borders where morphological identification may be challenging.
Species Identification	Enables the identification of fish species, especially during different life stages, using specific DNA markers.
Food Safety in Fisheries	Critical for assessing the quality of fish products, ensuring traceability, and preventing food piracy.
Biomonitoring and Ecological Assessment	Used to assess the presence of endangered species for conservation efforts and to monitor ecological indicators.
Cryptic Species Delimitation	Aids in identifying and delimiting cryptic fish species that may be challenging through traditional taxonomy.

Advantages of DNA Barcoding

- Widely used in fish identification.
- Generic primers available for COI barcode region COI is generally useful for distinguishing closely related and less closely related taxa.
- Alternate markers can be sequenced if COI barcode is not differential.
- May be useful for taxonomic analyses.

Disadvantages of DNA Barcoding

- Requires a large database of sequences for comparison.
- Prior knowledge of the barcoding region is required when applied diagnostically
- Since COI and other mitochondrial genes are maternally inherited, decreased barcode diversity may skew phylogenetic inferences.

Conclusion

DNA barcoding in fish species provides a powerful and efficient tool for species identification, contributing to various fields such as fisheries management, conservation, and ensuring the safety and authenticity of fish products in the market.

2. DNA Polymorphism

DNA marker technologies have revolutionized the way aquaculture genetics research is conducted. Widespread use of allozymes in the 1970s Genomics: Broadly defined as the study of genes and their functions, Genomics is rapidly impacting many facets of life, from health care and food safety to reproduction and law enforcement. Keys to the emergence of genomics were advances in DNA marker technology. Allozymes, mtDNA, RFLPs, RAPDs, AFLPs, microsatellites SNPs, and ESTs - potentially widespread utility in a variety of aquaculture endeavors.

Applications

- Assessment of genetic variability
- Inbreeding parentage assignment
- Species and strain identification
- Hybridization
- Marker-assisted identification of quantitative trait loci (QTL) through the construction of genetic linkage maps

Type I Versus Type Ii Markers and Polymorphic Information Content (PIC)

Molecular markers are classified into two categories: Type I are markers associated with genes of known function, while Type II markers are associated with anonymous genomic segments.

- Type I markers have utility in studies of:
- comparative genomics
- genome evolution
- candidate gene identification
- enhanced communication among laboratories.
- Comparative genomics deals with the similarity and differences found among genomes.

Type I markers serve as a bridge for the comparison and transfer of genomic information from a map rich species into a relatively map-poor species.

In general, type II markers such as RAPDs, microsatellites, and AFLPs are considered to be non-coding and therefore selectively neutral.

Such markers have found widespread use in population genetic studies whose characterizations of genetic diversity and divergence within and among populations are based on assumptions of Hardy–Weinberg equilibrium and selective neutrality of the markers employed.

Type II markers also have proven useful in aquaculture genetics for species, strain and hybrid identification, in breeding studies, and more recently as markers linked to QTL.

Types of DNA Polymorphism

- DNA polymorphism refers to variations in the DNA sequence among individuals or populations. These variations can arise from various processes, including mutation, recombination, and genetic drift. The study of DNA polymorphisms is essential for understanding genetic diversity, evolution, and disease susceptibility.
- Single Nucleotide Polymorphisms (SNPs): SNPs are the most common type of DNA polymorphism. They are single base-pair changes in the DNA sequence, occurring approximately once every 300 base pairs in the human genome. SNPs can be associated with diseases, traits, and drug responses.
- Insertions and Deletions (Indels): Indels involve the insertion or deletion of one or more nucleotides in the DNA sequence. They can cause frameshift mutations, altering the amino acid sequence of proteins.
- Copy Number Variations (CNVs): CNVs are DNA polymorphisms that involve changes in the number of copies of a particular DNA segment. They can affect gene dosage and gene expression levels, contributing to phenotypic diversity.
- Transposable Elements (TEs): TEs are sequences of DNA that can move from one location to another within the genome. They are a significant source of genetic variation and can influence gene expression and genome evolution.

Type I Markers

Type 1 markers are associated with gene of known function.

Types of Type 1 markers include

- Allozymes
- Isozymes
- Restriction fragment length polymorphism (RFLP)
- Expressed Sequenced Tag (EST)

Type II Markers

Type 2 Markers are associated with gene of unknown function

Types of Type II Markers are

- Randomly Amplified Polymorphic DNA
- Amplified Fragment Length DNA
- Single nucleotide Polymorphism
- Microsatellite Markers

Type I Markers

Allozyme markersss

	Allozymes	**Isozymes**
Definition	Allozymes are multiple forms of an enzyme that are coded by different alleles at same locus	Isozymes are multiple forms of an enzyme that are coded by different genes and different loci
Coded by	Different allele at the same locus	Gene at different loci
Number of genes involved	Single gene	Several different gene
Alleles	At the same locus	Different loci

Restriction Fragment Length Polymorphism (RFLP)

Restriction endonucleases are bacterial enzymes that recognize specific 4, 5, 6, or 8 base pair (bp) nucleotide sequences and cut DNA wherever these sequences are encountered, so that changes in the DNA sequence due to indels, base substitutions, or rearrangements involving the restriction sites can result in the gain, loss, or relocation of a restriction site.

Digestion of DNA with restriction enzymes results in fragments whose number and size can vary among individuals, populations, and species.

Traditionally, fragments were separated using Southern blot analysis (Southern, 1976), in which genomic DNA is digested, subjected to electrophoresis through an agarose gel, transferred to a membrane, and visualized by hybridization to specific probes.

Type II Markers

Amplified fragment length polymorphism

AFLP is a PCR-based, multi-locus fingerprinting technique that combines the strengths and overcomes the weaknesses of the RFLP and RAPD methods discussed previously.

Like RFLPs, the molecular basis of AFLP polymorphisms includes indels between restriction sites and base substitutions at restriction sites; like RAPDs, it also includes base substitutions at PCR primer binding sites.

The unique feature of the technique is the addition of adaptors of known sequence to DNA fragments generated by digestion of whole genomic DNA.

This allows for the subsequent PCR amplification of a subset of the total fragments for ease of separation by gel electrophoresis.

Its primary target of genetic variation is the same as RFLP, but instead of analyzing one locus at a time, it allows for the analysis of many loci simultaneously.

Randomly amplified Polymorphic DNA (RAPD)

RAPD procedures were first developed in 1990 using PCR to randomly amplify anonymous segments of nuclear DNA with an identical pair of primers 8–10 bp in length.

Because the primers are short and relatively low annealing temperatures (often 36–40 degree C) are used, the likelihood of amplifying multiple products is great, with each product (presumably) representing a different locus.

Because most of the nuclear genome in vertebrates is non-coding, it is presumed that most of the amplified loci will be selectively neutral.

Microsatellites markers

- ❑ Microsatellites consist of multiple copies of tandemly arranged simple sequence repeats (SSRs) that range in size from 1 to 6 base pairs (e.g., ACA or GATA)
- ❑ Abundant in all species studied to date, microsatellites have been estimated to occur as often as once every 10 kb in fishes (Wright, 1993).
- ❑ Microsatellites tend to be evenly distributed in the genome on all chromosomes and all regions of the chromosome. They have been found inside gene coding regions , introns, and in the non-gene sequences.
- ❑ The best known examples of microsatellites within coding regions are those causing genetic diseases in humans, such as the CAG repeats that encode polyglutamine tract, resulting in mental retardation.

- Most microsatellite loci are relatively small, ranging from a few to a few hundred repeats. The relatively small size of microsatellite loci is important for PCR-facilitated genotyping.
- Generally speaking, microsatellites containing a larger number of repeats are more polymorphic, though polymorphism has been observed in microsatellites with as few as five repeats.

Single Nucleotide Polymorphism (SNPs)

SNP is defined as a nucleotide site, for which a high substitution rate has been shown among individual samples in a population.

According to the more accurate definition given by Brookes, "SNPs are single base pair positions in genomic DNA, at which different sequence alternatives (alleles) exist in normal individuals in some population(s), herein the least frequent allele has an abundance of 1% or greater."

The restriction posed on frequency distinguishes SNPs from rare point mutations and implies the use of the former as genetic markers.

Based on this definition, SNP markers formally do not include single-nucleotide insertions/deletions (indels), though some authors combine these two polymorphism types with single-nucleotide substitutions under the common term of SNPs (simple nucleotide polymorphisms).

Molecular markers used in taxonomy

Tasks	Recommended marker system	Other useful marker types
Species identification	RAPD	AFLP, Microsatellites, isozymes
Strain identification	AFLP,Microsatellites	RAPD
Hybrid identification	RAPD	AFLP,Microsatellites,
Genetic resource/diversity analysis	AFLP, Microsatellites	RAPD

Significance of DNA Polymorphism

DNA polymorphisms are the genetic variations that make each individual unique. They are the foundation of genetic diversity within and among populations, and they play essential roles in various biological processes. Polymorphisms can affect gene function, protein structure, and ultimately, the phenotype of an organism. Understanding DNA polymorphisms is crucial for unraveling the genetic basis of traits and diseases in humans and other organisms.

3. Protein Analysis

Biochemical Genetic Markers: Principles and Types

Electrophoresis in Biochemical Genetic Markers

- Each fish species has different protein compositions at varying levels.
- Techniques that separate proteins aid in species identification, with electrophoresis being a crucial method.
- Proteins, composed of amino acids with charged side chains, have a net charge depending on the proportions of amino acids.
- Electrophoretic separation exploits the migration of proteins with different net charges and molecular sizes within an electric field.
- Proteins at their iso-electric point have a net charge of zero.

History and Basics of Electrophoresis

- Term "electrophoresis" means "transport by electricity."
- Known since the end of the 19th century.
- In 1807, Russian physicist Alexander Reuss observed colloidal particles moving towards the positive electrode when electricity passed through a tube containing water and clay.
- Electrophoresis describes the migration of charged particles under the influence of an electric field.

Theory of Movement in Electrophoresis

- When a potential difference is applied across electrodes, it generates a potential gradient (E).
- Charged particles migrate either to the cathode or anode based on their net charge.
- The force driving a charged molecule is the product of potential gradient and charge.
- Frictional force retards movement and depends on the molecule's size, shape, pore size of the medium, and buffer viscosity.
- Velocity (v) of a charged molecule in an electric field is given by $v = Eq/F$.

Charge and pH Influence

- Molecules exist in solution as charged particles at a given pH.
- Proteins have both acidic (negatively charged) and basic (positively charged) groupings, making them amphoteric molecules or zwitterions.
- Actual protein charge is the sum of all single charges.
- pH greatly influences total charge; at lower pH, migration is towards the cathode, and at higher pH, it's towards the anode.

- Ionic strength affects migration, with low ionic strength allowing higher rates.

Factors Affecting Migration in Electrophoresis

- The choice of buffer strength is crucial as it determines the applied electrical power.
- Rate of migration depends on the charge density (charge-to-mass ratio) of proteins; higher ratios lead to faster migration.

Biochemical Genetic Markers Types

- **Proteins:** Identification based on different protein compositions.
- **Amino Acids:** Variation in amino acid composition.
- **Enzymes:** Analysis of enzyme activity.
- **Nucleic Acids:** Detection of DNA or RNA variations.
- **Metabolites:** Measurement of metabolic products.

Gel Electrophoresis: Principles and Types

Introduction

Gel electrophoresis is a widely used technique in molecular biology and biochemistry for separating macromolecules, such as DNA, RNA, or proteins, based on their size, charge, or other physical properties. In gel electrophoresis, a gel (commonly agarose or polyacrylamide) serves as the support medium.

Types of Gel Electrophoresis

Based on Buffer System

Continuous Buffer System

- The simplest and most common method.
- The same buffer is used for electrical contact and gel soaking.
- Proteins migrate based on molecular size; electrical charge induces movement.

Discontinuous Buffer System

- Different buffers in electrode chambers and gel.
- Enhances protein concentration at the meeting point of the two buffers for improved resolution.

Based on Support Media

Supporting Media

- Inert, easy availability, high electrical conductivity, low adsorptivity, desirable sieving effect, controlled porosity, high transparency, low electroendosmosis, moderate to high rigidity, feasible preservation, low toxicity, and easy preparation are desired features.

- Various supporting media include cellulose acetate, agar, starch, and polyacrylamide.

Starch Gel Electrophoresis

- Starch used as support media due to its molecular sieving properties.
- Prepared by mixing hydrolyzed potato starch with an electrolyte buffer.
- Used for analyzing allozyme patterns.

Polyacrylamide Gel Electrophoresis (PAGE)

- Acrylamide monomer polymerized with a cross-linking agent, forming a gel with adjustable pore size.
- Native PAGE for separating proteins based on size and charge.
- Denaturing PAGE with SDS and mercaptoethanol to separate based on size alone.

Agarose Gel Electrophoresis

- Used for DNA and RNA separations.
- Agarose concentration (0.5-3%) influences mobility.
- Horizontal slab gels preferred for maximum mobility.

Cellulose Acetate Gel Electrophoresis

- Protein migration occurs in the buffer film on the gel surface.
- Medium has no influence on electrophoretic mobility.

Isoelectric Focusing (IEF)

- Uses polyacrylamide gel with polyamino, polycarboxylic acids with different pI values.
- Stable pH gradient along the gel in an electric field.
- Proteins migrate until reaching their isoelectric point.

Two-Dimensional (2D) Electrophoresis

- Combines isoelectric focusing and SDS-PAGE for enhanced protein separation.
- Resolves proteins based on charge and size in two independent dimensions.

Power Supply

DC battery or AC rectified to DC.

Constant voltage, current, or power supplied by electrophoresis power supply.

Buffers

- Solutions of weak acids and their salts to maintain constant pH.
- Common buffers include phosphate, Tris-Borate-EDTA (TBE), Tris-Acetate EDTA (TAE), Tris Glycine (TG), Tris-Citrate-EDTA (TCE), Tris-EDTA (TE), Tris-Maleic acid-EDTA (TME), and Lithium Borate (LB).

Sources and Extraction of Proteins: Principles and Applications

Introduction

- For electrophoretic separation, proteins need to be in solution. Various body fluids such as plasma, serum, milt, haemolymph, or aqueous extracts of tissues like muscle, liver, and eyes are common protein sources. Extracts are typically made using distilled water, 10% sucrose solution, or specific extraction buffers. The pH and ionic concentration of the buffer influence the differential extraction of proteins.

Storage and Denaturation:

- After extraction, proteins start to denature rapidly, necessitating immediate use or deep-freezing for preservation. Storage at ultra-low temperatures like -196°C or -80°C retains protein activity, while -40°C is suitable for short-term storage. However, denaturation is relatively rapid at -18°C, leading to altered electrophoretic patterns.

Types of Proteins for Electrophoresis

Sarcoplasmic Proteins (Water-soluble Proteins)

- Soluble proteins in the sarcoplasm within the sarcolemma.
- Include albumins, myogens, and glycolytic enzymes.
- Genetic differences are pronounced, aiding species identification.

Myofibrillar Proteins (Salt-soluble Proteins)

- Found in myofibrils of muscle fibers.
- Myosin and tropomyosin are used in fish species identification via electrophoresis.
- Electrophoretic patterns of myosin and tropomyosin aid in species differentiation.

Eye-lens Proteins

- Synthesized by a single cell type in the eye lens.
- Alpha, beta, and gamma crystallines are distinguishable by electrophoretic and immunological techniques.
- Used in taxonomic studies; alpha-crystallin is organ-specific, while beta and gamma crystallines are species-specific.

Allozymes

- Functionally similar forms of enzymes encoded by one or more loci.
- Allozymes are products of different alleles at the same locus.
- Used as genetic markers for stock identification and population divergence studies.

Staining Systems for Protein Detection

- Once electrophoresis is done, staining is essential to identify protein positions. Common stains include Ponceau red S, Amido black 10 B, Kenacid blue, or Coomassie Brilliant Blue R. Specific staining techniques involve detecting proteins based on their functional properties.

Technique	Principle	Application	Example
Direct Substrate to Product Conversion	The enzyme converts the substrate directly into a visible product.	Identifying enzymes with observable products.	Detection of enzymes catalyzing reactions with readily observable products.
Substrate to Non-visible Product	The enzyme converts the substrate into a non-visible product, made visible by additional histochemicals.	Identifying enzymes with non-visible products that become visible after treatment.	Staining enzymes where the product is not directly observable.
Substrate to Non-visible Product with Linking Enzyme	The enzyme converts the substrate into a non-visible product, and a linking enzyme transforms it into a visible product.	Identifying enzymes indirectly through a linking enzyme.	Staining systems where an intermediary reaction is involved before the final product becomes visible.
Dehydrogenase Reactions	Identify reactions involving hydrogen ion transfers using NAD or NADP, visualizing with tetrazolium salts.	Detection of dehydrogenases in various metabolic pathways.	Detection of lactate dehydrogenase or alcohol dehydrogenase activities.
Specific Enzyme Staining	Use specific substrates for targeted enzymes along with histochemical stains.	Identification of enzymes with known substrates and reactions.	Detection of alkaline phosphatase using specific substrates producing colored or fluorescent products.
Isozyme and Allozyme Identification	Identify specific isozymes or allozymes based on their characteristic enzymatic activities.	Differentiate between variants of enzymes encoded by different alleles.	Allozyme electrophoresis to study genetic markers in populations.

Conclusion

- Electrophoresis, a fundamental biochemical method, plays a central role in molecular-based studies, especially in genetics. It allows for the separation and visualization of macromolecules, aiding in genetic diversity studies, taxonomic identification, and conservation efforts. Staining techniques, particularly enzyme staining, enhance the specificity and functionality of protein detection, contributing to a better understanding of genetic markers and population dynamics.

Glossary Terms

Acanthopterygii: A superorder of ray-finned fish, including many common marine species.

Anabantoidei: A suborder of labyrinth fish known for their ability to breathe air.

Anal Fin: The fin located on the ventral side near the anus in fish.

Atheriniformes: An order of fish that includes silversides and rainbowfish.

Amphioxus: A small, fish-like marine invertebrate often used in studies of vertebrate evolution.

Batoidea: A subclass of cartilaginous fish that includes rays, skates, and sawfish.

Branchiostegal Rays: Bony rays supporting the gill membranes in fish.

Buccal Cavity: The oral cavity or mouth of a fish.

Bony Fishes (Osteichthyes): A class of fish characterized by bony skeletons rather than cartilaginous ones.

Biogeography: The study of the distribution of species and ecosystems in geographic space and through geological time.

Caudal Peduncle: The narrow part of a fish's body to which the caudal or tail fin is attached.

Chondrichthyes: A class of cartilaginous fish, including sharks, rays, and chimaeras.

Cichlidae: A family of freshwater fish known for their diversity and popularity in the aquarium trade.

Cladogram: A diagram depicting patterns of shared characteristics among species.

Cleithrum: A bone in the pectoral girdle of fish, supporting the base of the pectoral fin.

Dentary Bone: The lower jawbone in fish, forming part of the mandible.

Dipnoi: A class of fish that includes lungfish, capable of breathing air.

Dorsal Fin: The fin located on the back of a fish.

Dioecious: Having separate sexes, with individuals being either male or female.

Dermal Denticles: Small, tooth-like structures on the skin of cartilaginous fish.

Endemic: Native and restricted to a specific geographic region.

Elopomorpha: An order of fish that includes eels and tarpons.

Euryhaline: Fish capable of tolerating a wide range of salinity in water.

Ectothermic: Cold-blooded; organisms that rely on external sources of heat to regulate body temperature.

Epibenthic: Organisms that live on the surface of the seafloor.

Fusiform: A streamlined and spindle-shaped body form that reduces drag in fast-swimming fish.

Foramen Magnum: An opening in the skull through which the spinal cord passes.

Fusimotor Neurons: Neurons controlling muscle spindle activity, important in fish movement.

Fish Base: A comprehensive database of fish species, their taxonomy, and biological information.

Fringed Scale: Scales with fringed edges, common in some fish species.

Ganoid Scales: Bony, diamond-shaped scales found in some primitive fish.

Gill Arch: The skeletal structure supporting the gills in fish.

Gonopodium: Modified anal fin in some fish species, adapted for copulation.

Gastric Mill: A structure in the stomach of some fish that aids in grinding food.

Genomic Evolution: The study of changes in the structure and organization of an organism's genetic material.

Homocercal: A type of tail in which the upper and lower lobes are of equal size.

Heterocercal: A type of tail where the upper lobe is larger than the lower lobe.

Hydrodynamics: The study of fluid flow, crucial in understanding fish locomotion.

Hemibranch: Half of a gill arch, often used in respiration.

Haplotype: A set of DNA variations or markers that tend to be inherited together.

Ichthyology: The branch of zoology that deals with the study of fish.

Infraclass: A taxonomic rank below class and above superorder.

Infauna: Organisms that live within the substrate at the bottom of aquatic environments.

Intermaxillary Bone: A bone forming part of the upper jaw in fish.

Interspecific: Relating to interactions between different species.

Jugular Pelvic Fin: Pelvic fins located beneath the pectoral fins in some fish.

Jawless Fish: Fish lacking jaws, such as lampreys and hagfish.

Juvenile: The stage of development following larval or fry stage, but before adulthood.

Johannesson, Kerstin: A biologist known for her work on speciation and adaptation in marine snails.

Joinvillea: A genus of deep-sea fish in the family Stomiidae.

Keystone Species: A species that has a disproportionate impact on its ecosystem.

Kinetosome: A microtubule-organizing center in certain cells, important in the study of ciliary movement in fish.

Kingdom: The highest taxonomic rank in the biological classification.

Koi: A type of domesticated carp often kept for ornamental purposes.

K-strategist: Species that produce fewer offspring but invest more in their care.

Lateral Line System: A sensory system detecting changes in water pressure and vibration.

Leptocephalus: The larval form of eels, characterized by a flat and transparent body.

Linnaeus, Carl: A Swedish botanist who laid the foundations for the modern biological naming scheme of binomial nomenclature.

Lamprey: A jawless fish known for its parasitic feeding habits.

Lobe-finned Fish (Sarcopterygii): A class of fish characterized by lobed pectoral and pelvic fins.

Myomere: A segmental muscle structure in fish, responsible for lateral movement.

Monophyletic Group: A group of organisms that includes an ancestral species and all its descendants.

Migratory: Relating to the regular movement of fish from one region to another.

Maxilla: A bone forming part of the upper jaw in fish.

Marine Protected Area (MPA): A designated area where human activities are managed to protect the marine environment.

Nasal Pit: Depression or cavity in the head of some fish, housing the nostrils.

Nematocyst: A specialized stinging cell found in some fish and other organisms.

Notochord: A flexible rod-like structure present in the embryos of all chordates, including fish.

Neoteny: The retention of juvenile characteristics into adulthood.

Neuston: Organisms living at the air-water interface.

Otolith: A calcareous structure in the inner ear of fish, used for balance and hearing.

Oviparous: Reproducing by laying eggs.

Ovipositor: A tube through which fish and other animals deposit eggs.

Osteostraci: An extinct class of jawless fish with bony armor.

Omnivorous: Consuming both plant and animal matter.

Phylogenetics: The study of evolutionary relationships among organisms.

Pelagic: Relating to fish that live in the open ocean rather than near the coast.

Pinnate Fin: A fin with feather-like projections, found in some fish species.

Pterygiophore: Bony or cartilaginous support for a fin in fish.

Polyandry: A mating system in which a female mates with multiple males.

Quantitative Genetics: The study of the inheritance of traits with a genetic basis.

Quillback: A type of freshwater fish in the sucker family.

Quadrate Bone: A bone in the skull of fish, forming part of the jaw joint.

Quadrivium: A method of classification based on four main criteria.

Queriman: A genus of deep-sea fish in the family Stomiidae.

Ray-finned Fish (Actinopterygii): A class of fish with fins supported by bony rays.

Red List: A list of species at risk of extinction, maintained by the International Union for Conservation of Nature (IUCN).

Roe: Fish eggs, especially from a female fish.

Rheotaxis: The orientation or movement of an organism in response to water current.

Rhomboid Scales: Diamond-shaped scales found in some fish species.

Sarcopterygii: A subclass of lobe-finned fish, including lungfish and coelacanths.

Sympatric Speciation: The evolution of new species within the same geographical area.

Supraorbital Canal: A sensory canal running along the top of the head in some fish.

Superorder: A taxonomic rank below class and above order.

Siphonophores: Gelatinous colonial organisms often mistaken for jellyfish.

Trophic Level: The position of a species in a food chain or web.

Teleost Fish: A subclass of bony fish with well-developed jaw bones.

Threatened Species: A species at risk of becoming endangered, as per the IUCN Red List.

Telson: The posterior, unsegmented part of the body in some fish.

Transverse Scale: Scales arranged in a transverse (horizontal) pattern on a fish's body.

Ultraviolet Vision: Some fish can perceive ultraviolet light, which is invisible to humans.

Underfishing: Fishing at a level that allows the fish population to maintain a healthy size.

Umbilical Cord: A structure connecting the developing fish embryo to the yolk sac.

Ungulate: A hooved mammal, but sometimes used to describe the caudal fin of some fish.

Unisexual Reproduction: Reproduction involving only one sex, as seen in some fish.

Ventral: The lower surface of a fish.

Viviparous: Giving birth to live young rather than laying eggs.

Vertebral Column: The backbone or spine of a fish.

Vomer: A bone in the skull of fish, forming part of the upper jaw.

Viviparous: A reproductive strategy where the offspring develop inside the mother and are born alive.

Wrasse: A family of marine fish known for their vibrant colors.

Whisker: A sensory organ found in some fish, especially catfish.

White Muscle: A type of muscle tissue in fish associated with bursts of fast swimming.

Wetland: A habitat characterized by saturated or seasonally inundated soil.

Wooden Tongue: A disorder in fish where the tongue becomes discolored or deteriorates.

Xanthophore: Pigment cells responsible for yellow and orange coloration in fish.

References

1. Bal, D.V. and K.V. Rao , 1990. Marine fisheries of India. Tata McGraw-Hill Publishing Company Limited, New Delhi. p. 472.
2. Biswas, S.P. (2002). Fundamentals of Ichthyology Narendra Publishing House, Delhi. 392 pp.
3. Day, F. 1878 The fishes of India, being a natural history of the fishes known to inhabit the seas and freshwaters of India, Burma and Ceylon, Part 4: i-xx + 553-779 p. + pls. 139 - 195.
4. Eschmayer W. (Ed) 1998. The Catalog of Fishes. San Francisco: California Academy of Sciences.
5. Fischer, W. and G. Bianchi (Eds.) 1984. FAO species identification sheets for fishery purposes. Western Indian Ocean (Fishing Area 51). Prepared and printed with the support of the Danish International Development Agency (DANIDA). Rome, Food and Agricultural Organization of the United Nations, Vols. 1-6.
6. Ian, S.R., Munro, 1982. The marine and freshwater fishes of Ceylon. Narendra Publishing House, New Delhi. pp. 1-351.
7. Jayaram, K.C. (1999): The freshwater fishes of the Indian region. Narendra Publishing House, Delhi. 551 pp.
8. Jayaraman, K.C. (2002). Fundamentals of Fish Taxonomy. Narendra Publishing House. Delhi, 174. pp.
9. 9. Nelson, J.S., 1976. Fishes of the world. A Wiley - Inter science publication. JohnWiley & Sons (N.Y) pp. ix + 1- 416.

10. Ramaiyan, V. and R. Sivakumar, 1991. Sharks, skates and rays (an aid to the identification of sharks, skates and rays of Parangipettai Coast). Centre of Advanced Study in Marine Biology (Annamalai University), Parangipettai, India. 55 pp.

11. Talwar, P.K. and A. G. Jhingran 1991 Inland Fishes of India and adjacent countries. Vol.1, IBH Publishing, New Delhi.

12. Venkatarmani, V.K., P. Jawahar, R. Santhanam and T. Vaitheeswaran, 2004. Marine Ornamental Fishes of Gulf of Mannar. Monograph (ICAR/NATP/CGP), FC & RI, TANUVAS. 176 pp.

13. Venkataramanujam, K., V.K. Venkatramani and N. Ramanathan ,1996. A manual of finfish Taxonomy. Fisheries College and Research Institute. 80 pp.

14. Fish base, 2023. Retrieved from website.

15. Worms, 2023. Retrieved from website.

16. R. Durairaja, S. Aruna and S. Felix, 2016. Manual on Taxonomy of Fin Fish Fisheries College and Research Institute TAMIL NADU FISHERIES UNIVERSITY Ponneri, Thiruvallur District.

17. FAO Fish Identissfication manuals (Western and Eastern Indian Oceans)

18. "Textbook of Fish Biology and Fisheries" by S.S. Khanna and H.R. Singh.

9 789359 195704

Printed at Repro India Ltd.